AF862402

Seed Dispersal by Ants in a Deciduous Forest Ecosystem

Seed Dispersal by Ants in a Deciduous Forest Ecosystem

Mechanisms, Strategies, Adaptations

by

Elena Gorb
National Taras Shevchenko University of Kyiv, Ukraine

and

Stanislav Gorb
Schmalhausen Institute of Zoology, Kyiv, Ukraine

KLUWER ACADEMIC PUBLISHERS
DORDRECHT / BOSTON / LONDON

A C.I.P. Catalogue record for this book is available from the Library of Congress.

ISBN 1-4020-1379-5

Published by Kluwer Academic Publishers,
P.O. Box 17, 3300 AA Dordrecht, The Netherlands.

Sold and distributed in North, Central and South America
by Kluwer Academic Publishers,
101 Philip Drive, Norwell, MA 02061, U.S.A.

In all other countries, sold and distributed
by Kluwer Academic Publishers,
P.O. Box 322, 3300 AH Dordrecht, The Netherlands.

Printed on acid-free paper

All Rights Reserved
© 2003 Kluwer Academic Publishers
No part of this work may be reproduced, stored in a retrieval system, or transmitted
in any form or by any means, electronic, mechanical, photocopying, microfilming, recording
or otherwise, without written permission from the Publisher, with the exception
of any material supplied specifically for the purpose of being entered
and executed on a computer system, for exclusive use by the purchaser of the work.

Printed in the Netherlands.

Contents

Foreword

In many nations, there is a fairy-tale about assiduous ants, which help a poor hard-working stepdaughter to sort out seeds. This idea about the relationship between ants and seeds is based on the knowledge about the immense invisible work performed by ants around the clock in fields and forests. Countless ants transport and deposit seeds and thereby influence the survival, death, and evolution of many plant species. In the evolution of plants, seed dispersal by ants (myrmecochory) has independently appeared many times in different lineages. More than 3000 plant species use the help of ants to be planted, and this is not the final number.

Myrmecochory is a very interesting and rather enigmatic form of mutualistic ant-plant interrelations. This phenomenon is extremely complex, because of hundreds of ant species connected with hundreds of plant species. It is always difficult to predict fate for plants' descendants, but this becomes even more difficult, if such active carriers as ants take part in seed dispersal. Mutualism is only then stable, when both partners profit from such interrelation, and do not cheat. Plants that interact with ants are furnished with morphological and physiological adaptations, whereas their agile partners possess specific behavioural features promoting dispersal of plant diaspores. Do plants really benefit from interactions with various ant species? What are the intimate mechanisms of ant-plant interrelations in the temperate deciduous forest? How does seed dispersal by ants effect seedling survival? What is the role of behavioural factors, such as ant community interactions, agility and spatial orientation of different ant species in seed dispersal? To answer these and other questions, the authors have effectively combined a thorough approach to investigating plants' morphological and physiological adaptations with elegant field experiments on ants' behaviour. This monograph is a first attempt to collect information about morphology, ecology, and phenology of ants and plants from one ecosystem. This tremendous amount of work is supplemented by experimental data clarifying interactions between particular species and groups of species.

Formica polyctena was selected by the authors for many experiments as a focal species. It is a good idea to study the role of workers of this species as seeds carriers, because *Formica* s. str. ants have large territories and often dominate in the ecosystem of the deciduous forest. From my own experience, I know these ants as highly social and territorial species possessing, probably, the most sophisticated and reliable communicative system among ants. Additionally, workers of *Formica* s. str. interact with other ant species in a complex way that includes different components ranging from predation and competition to cooperation and learning from other species by observation (*Reznikova, 1982, Behaviour, 80: 84-95*). Numerous behavioural observations, made by the authors, are very important in the estimation of possible consequences of particular behaviour of ants for individual plants and for the forest vegetation.

The authors give readers a panoramic view of the hidden, poorly-known interrelations not only between pairs of ants and plant species, but also between species communities in the ecosystem. Knowledge of ecological networks is of general interest, because it aids in the understanding of ecosystem dynamics. The authors considered not just one aspect of animal-plant relationships, but tried to show them in all their complexity. Some aspects of the ant-plant interactions, described in the book, may be of interest to botanists, others again to zoologists or ecologists, but the entire work is a splendid example of a marriage of these biological disciplines.

Finally, the authors have put together an excellent summary of experimental methods in a separate chapter at the end of the book. This will be a great assistance to future generations of students beginning to study ant-plant interactions. This well-written and richly-illustrated book will, without doubt, encourage others to follow in collecting new seeds for the nest of biological knowledge.

Zhanna I. Reznikova
Professor of Zoology
Novosibirsk State University, Russia

Volodymyr I. Czopyk
Professor of Botany
Kyiv National University, Ukraine

Preface

Dispersal of diaspores (seeds and fruits) is an important phase in the life cycle of flowering plants. It determines dynamics of the natural habitats of species, influences formation, structure, and succession of the vegetation, and promotes cross-pollination. Knowledge about diaspore dispersal methods is a necessary condition for the understanding of processes and mechanisms of species survival in evolution, their adaptations to environmental changes. This knowledge is also important to predict population dynamics and potential changes of the natural habitat in particular species.

One of the wide variety of plant dispersal methods is myrmecochory or diaspore dispersal by ants. This dispersal method is registered in more than 3000 plant species from 67 families. Myrmecochorous species compose an important component of the floras not only of tropical rainforests, but also temperate areas of Europe. In some ecosystems, they may compose up to 40–50% of herbaceous plant species. Taking into account the importance of myrmecochory for ecology, agriculture, and forestry, the study of many aspects of the mechanisms of seed dispersal by ants seems to be neglected.

Myrmecochory in eastern European forests, especially in the Ukraine, is poorly studied. There are only a few purely descriptive publications devoted to this topic (Novikov, 1939; Levina, 1957; 1967; Targonsky, 1995). With regard to a number of herbaceous species employing this dispersal method in European deciduous forests, the authors spent five field seasons in one selected ecosystem in the middle Dnipro area (50°17´ North Latitude, 31°31´ East Longitude) in order to clarify mechanisms of ant-dispersal of plant diaspores. Principal results of the field experiments and laboratory studies were partly published in ecological and botanical journals (Gorb, 1998; 2000; Gorb and Gorb, 1995a; b; c; 1996; 1997; 1998a; b; 1999a; b; 2000; 2001a; b; Gorb et al., 1996; 1997; 2000; Sindarovskaya et al., 1997). These studies and the present book was an enthusiastic project of the authors. In the book, we tried to integrate our data with existing knowledge to give the reader an impression of the complexity of the phenomenon of myrmecochory even in such a «species-poor» habitat, as a temperate deciduous forest. In a much broader aspect, diaspore dispersal by ants may be an interesting model system for studies of questions of the co-evolution between insects and flowering plants.

This book is a complex study of myrmecochory in one ecosystem. It supports the hypothesis that a complex of ant species influences a complex of plant species through seed and fruit dispersal. Some characteristic aspects of the ant-plant interactions in cases of obligate and facultative myrmecochory are described here. It is shown that myrmecochorous plants have numerous adaptations to diaspore dispersal by ants (so-called myrmecochorous syndrome) (**chapter 1**). Three main stages of myrmecochory (removal, transport, and secondary relocation of diaspores) as well as factors influencing removal and transport of seeds are studied (**chapters 2 – 7**).

It is shown that diaspores of different plant species are removed by ants at different rates (**chapter 2**). Morphological, behavioural and ecological factors, influencing diaspore removal rates, are analysed in a series of field experiments (**chapter 3**). It is proven that diaspore attractiveness increases with an increase of the diaspore size and only weakly correlates with the elaiosome size. The effect of the ant species composition of a microsite on the diaspore removal is clarified (**chapter 4**). It is shown that, in forest microsites with different complexes of ant species, foragers of the dominating ant species mainly contribute to the visiting and removal rates of diaspore depots.

In field experiments, the proportion of diaspores dropped during transport by ants is estimated for some plant species (**chapter 5**). It is revealed that the amount of dropped diaspores is plant-species-specific. Principal characters influencing dropping rates are dimensions of the diaspore as well as size of ant individuals. Dropping rate increases with a decrease of diaspore dimension and inversely correlates with the removal rate.

In **chapters 6** and **7,** the seed flow in ant species territories is analysed. It is revealed that diaspores of myrmecochores prevail in the seed flow directed to the ant nest. The significance of secondary relocation for an increase of dispersal distance is shown (**chapter 8**). It is also demonstrated how nest-cleaning behaviour of different ant species influences spatial distribution of myrmecochores. Two computer models are proposed. The first one describes the effect of seed producing capacity and seed dropping rates during ant transport on the dispersal distance of diaspores. The dependence of diaspore distribution on the presence of secondary relocation of diaspores is shown in the second model. If the secondary relocation is present, distances between neighbouring seedlings and between seedlings and the parent plant increases considerably.

A particular plant species, ant species, and microsite are taken into account, while analysing selective advantages of myrmecochory (**chapter 9**). The relevance of some hypotheses about selective advantages of myrmecochory is tested for a few myrmecochores. The interrelations between non-myrmecochorous plants and ants are discussed (**chapter 10**). In **chapter 11,** the ecosystem, in which experiments were carried out, is described. Those who wish to repeat some experiments or to start to study this fascinating phenomenon on one's own will find detailed descriptions of experimental approaches and some useful advise on experimental design.

At the end of each chapter, a short summary is provided to emphasise principal statements of the chapter. A short annotated list of terms at the end of the book may aid in a better understanding of botanical and entomological terminology used.

Since the topic of the book crosses different biological disciplines, it would be impossible without the additional expertise of botanists, zoologists, ecologists. We are thankful to V.I. Czopyk and N.N. Bortnjak (Department of Botany, Kyiv National University, Ukraine) for their continuous support. A.G. Radchenko (Schmalhausen Institute of Zoology, Kyiv, Ukraine) provided ant species determination. N.M. Panasenko, J.R. Sindarovskaja, E.S. Neizvestnaja (Department of Botany, Kyiv National University, Ukraine) helped in field experiments. We wish to thank H. Schwarz and J. Berger (Max-Planck-Institute of Developmental Biology, Tübingen, Germany) for their kind help with diverse microscopical procedures. Discussions about ecological networks of the deciduous forest with L. Vinogradskaja (Department of Zoology, Kyiv National University, Ukraine) are also greatly acknowledged.

We would particularly like to acknowledge the help of those colleagues who have commented on the book (N. Rowe, University of Montpellier, France; V.I. Czopyk, L.F. Kucherjava, Kyiv National University, Ukraine; L.I. Frantsevich, A.I. Kotenko, A.G. Radchenko, V.G. Radchenko, Schmalhausen Institute of Zoology, Kyiv, Ukraine; S.V. Morozjuk, Department of Biology and Geography, Dragomanov Pedagogical University, Kyiv, Ukraine; S.L. Mosjakin, Cholodny Institute of Botany, Kyiv, Ukraine; Z.I. Reznikova, Novosibirsk University, Russia; V.G. Sobko, V.K. Gorb, Central Botanical Garden of the Academy of Sciences of the Ukraine, Kyiv). N. Rowe and V. Kastner provided numerous linguistic suggestions and corrections that improved the text.

Elena and Stanislav Gorb

Introduction: an historical background

The fact that ants collect and transport plant diaspores has been known since ancient times. The first scientific reports about diaspore dispersal by ants concern tropical regions (Sykes, 1834; Buckley, 1860) and gradually information about ants transporting seeds accumulated together with data on the plant species using ant behaviour for relocation of their seeds and fruits (Lincecum, 1862; Moggridge, 1873; Kuntze, 1876; Andre, 1881; Adlerz, 1886). A variety of suggestions have been expressed to explain the motives impelling ants to remove and transport diaspores (Lundstroem, 1887; Lubbock, 1887; Ludwig, 1898; 1899). Some authors have indicated the importance of diaspore appendages and their fats for attracting potential dispersal agents (Ruppert, 1893; Robertson, 1897; Kerner, 1896; 1898; Lagerheim, 1900).

The first complex study of European myrmecochores was carried out by R. Sernander (1901; 1906), who secured the use of the terms «myrmecochory», «myrmecochore», and proposed a new term «elaiosome» for fat-containing appendages of the seeds and fruits which attract ants. He worked out the first classification of myrmecochorous plants. It was based on elaiosome morphology (morphological aspect), type of embryo (anatomical aspect), and contribution of the myrmecochory to diaspore dispersal (ecological aspect). This classification was further modified (Ridley, 1930; Levina, 1957) or expanded (Berg, 1975; Drake, 1981) by later authors.

Experiments by Sernander drew the attention of many scientists to problems of myrmecochory and ant-plant relationships in general. Original observations and numerous data from the literature were accumulated in [illegible]ographs devoted to the methods of diaspore dispersal in general and to [illegible]chory in particular (Wheeler et al., 1921–1922; Ridley, 1910; 1930;

Ulbrich, 1919; 1928; Levina, 1957; 1967; Pijl, 1972). The division of plants into obligate and facultative myrmecochores (see chapters 1, 2) was proposed. Ulbrich (1919; 1928) first indicated that myrmecochorous plants possess not just elaiosomes, but a broad complex of adaptations to ant dispersal, later called the «myrmecochorous syndrome». The significance of the syndrome for ant attraction and for the dispersal mechanism in general was actively discussed.

Since the 1930's, a number of studies have been done, in which anatomical features of elaiosomes were described (Baciu, 1931; 1935; Nordhagen, 1932; 1959; Szemes, 1943; Geitler, 1944; 1949; 1950). Attention was paid mainly to the elaiosome design, origin, and chemical content (Bresinsky, 1963; Komar, 1965;1973; 1974; 1978; Roth, 1977).

A series of studies on the morphology, development, and dispersal of diaspores of a number of Australian plant species was published in the 1950's to 1970's by R. Berg (1954; 1958; 1966; 1969; 1972). Later, Australian myrmecochores were actively studied by D. Davidson and S. Morton (1981a; b) and by B. Rice and M. Westoby (Rice and Westoby, 1981; Westoby and Rice, 1981; Westoby et al., 1982). In publications by S. Handel, the dissemination ecology of some myrmecochorous plant species was studied (Handel, 1976; 1978a), and the role of myrmecochory in North American forest ecosystems (Handel, 1978b; Handel et al., 1981) was analysed.

An important contribution for understanding myrmecochory was made by the works of A. Beattie who, in co-operation with other investigators, did a lot of fascinating studies devoted to various aspects of this phenomenon (Beattie and Lyons, 1975; Pudlo et al., 1980; Beattie and Culver, 1981; 1982; 1983; Culver and Beattie, 1983). Results of these studies were brought together in review on the distribution of myrmecochory throughout the world (Beattie, 1983). In the review, the author listed plant genera containing myrmecochorous species, and systematically analysed groups and geographical distributions of ant genera with species reported as being seed and fruit dispersers. Additionally, he proposed and discussed five hypotheses about selective advantages of myrmecochory. This work, up to now, remains the most general overview of the phenomenon.

A. Beattie's review stimulated numerous studies on myrmecochory published in the last two decades. The majority of these studies recorded new myrmecochorous plants or catalogued myrmecochores of particular climatic regions (Clay, 1983; Davidson and Morton, 1984; Abbott and Heurck, 1985; Roberts and Heithaus, 1986; Bregman, 1988; Bullock, 1989; Smith, 1989; Horvitz, 1991; Bond et al., 1991; Targonsky, 1995). Some studies provided new details on anatomy (Tahtadzhjan, 1985; 1988; 1991; 1992; Werker, 1997) and chemical composition of elaiosomes (Hocking ar

Kortt, 1987; Skidmore and Heithaus, 1988; Brew et al., 1989; Hughes et al., 1994). In field experiments, details of dispersal mechanisms and selective advantages of myrmecochory were tested (Kjellsson, 1985; Auld, 1986; Horvitz and Schemske, 1986b; Gunther and Lanza, 1989; Oostermeijer, 1989; Hughes and Westoby, 1990; 1992a; Affre et al., 1995; Corff and Horvitz, 1995; Gorb and Gorb, 1995a; b; 1998a; 1999a; b; 2000; Mark and Olesen, 1996; Passoss and Ferreira, 1996; Gorb, 1998). Theoretical investigations aided in integrating a huge amount of data and in extracting general principles of diaspore dispersal by ants (Heithaus et al., 1980; Howe and Smallwood, 1982; Cushman and Beattie, 1991).

Fig. *Formica polyctena* worker carrying a seed of *Viola odorata*.

Thus, owing to a great number of publications on various aspects of myrmecochory, the general scenario of this type of diaspore dispersal is relatively clear: (1) elaiosome-bearing diaspores are removed and transported by ants to their nests (Fig.); (2) in the nests, the ants consume the energetically-valuable elaiosomes and (3) throw away the diaspore bodies (diaspores without elaiosomes), which are capable of germination. In spite of numerous studies on plant adaptations with this method of diaspore dispersal, the influence of different factors on diaspore removal, transport, dropping, and seedling survival remains unclear. Seed flow and secondary relocation of diaspores in ant territories have not been previously studied. Depending on experiments of different authors, data on selective advantages of myrmecochory are often contradictory.

As this introductory chapter shows, a large amount of literature data on myrmecochory does exist. However, the data were obtained by different authors, in different ecosystems, on different species of plants and ants or ant/plant species combinations. Sometimes, it is very difficult to compare and integrate such data. The principal advantage of this book is that it is based on the authors' original data of field experiments and laboratory studies; and that it covers a variety of little-known aspects of myrmecochory within the same ecosystem of a temperate deciduous forest.

SUMMARY

- A short literature review on the phenomenon of myrmecochory is presented.
- The general scenario of this method of diaspore dispersal is as follows: elaiosome-bearing seeds are removed and transported by ants to their nests; ants consume energetically valuable elaiosomes and throw away the rest of the diaspore.
- In spite of numerous studies on plant adaptations to this method of dispersal, the influence of different factors on seed removal, transport, dropping, and seedling survival remains unclear. Diaspore flow and secondary relocation of diaspores in ant territories have not been previously studied.
- Data on selective advantages of myrmecochory were obtained by previous authors in different ecosystems. Therefore, it is difficult to integrate these results and extrapolate them to other ecosystems. Based on the authors' original experimental studies and literature data, this book covers a variety of little-known aspects of myrmecochory in the ecosystem of a temperate deciduous forest.

Chapter 1

The myrmecochorous syndrome

Many flowering plants have evolved morphological, anatomical, biochemical, phenological adaptations, amongst other, for promoting an effective diaspore dispersal. Some adaptations appeared in plant evolution as part of a diaspore dispersal mechanism by ants. This introductory chapter discusses a variety of plant adaptations for diaspore dispersal by ants and analyses the significance of such adaptations for ant attraction. We combine literature data with results of our own field experiments.

1. WHAT IS MYRMECOCHORY?

Diaspore dispersal by animals is called *zoochory*. *Myrmecochory* is one such mechanism, where ants serve as dispersal agents. In the case of specialised myrmecochory (Fig. 1.1), diaspores have specialised fat-containing appendages, *elaiosomes*, which attract ants and lead to release their carrying behaviour (Sernander, 1906). Ant workers take such diaspores and transport them to their nests. There, ants remove the elaiosomes from the diaspore bodies (diaspores without elaiosomes) and throw the otherwise undamaged diaspore bodies out of the nest. The abandoned diaspore bodies maintain their ability to germinate (Handel, 1976; 1978; Kjellsson, 1985; Auld, 1986). During evolution, true mutualism between myrmecochorous plants and ants evolved: ants consume the energetically-valuable elaiosomes and plants take advantage of the diaspore dispersal by the ants (Heithaus et al., 1980; Cushman and Beattie, 1991). In the case of non-specialised myrmecochory, diaspores have no elaiosomes. Harvesting ants (e.g. representatives of the genus *Messor*) usually collect such diaspores, deposit them into their nests and consume the entire seeds. Occasionally, some seeds get dropped during transport, and others are deposited in underground granaries. These seeds have the chance to germinate depending on the depth of the location (Ridley, 1930; Pijl, 1982). Since in the case of elaiosomeless seeds and harvesting ants, the number of consumed seeds largely surpasses the number of lost or forgotten seeds, the phenomenon of diaspore dispersal by harvesting ants does not result from the true mutualism but merely from a side effect of ants gathering seeds for food (Handel and Beattie, 1990). This chapter reviews diverse plant adaptations to the specialised myrmecochory.

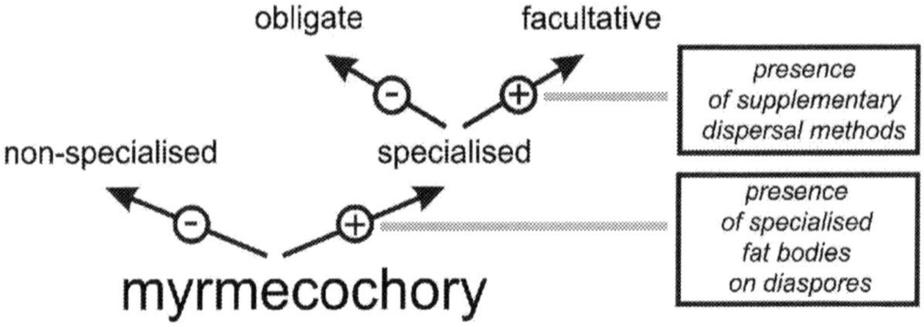

Fig. 1.1. Diagram of different types of myrmecochory.

2. THE CONCEPT OF THE MYRMECOCHOROUS SYNDROME

Seed dispersal, which does not necessarily require dispersal agents, is called *autochory*. When dispersal agents, such as wind, water, animals, or humans, are involved, seed dispersal is known as *allochory* (Levina, 1957). Both dispersal methods are reflected in particular plant adaptations. The complex of various adaptations to the particular dispersal method is called a *dispersal syndrome* (Pijl, 1982). Myrmecochorous plants usually possess special morphological, anatomical, biochemical, and phenological adaptations, enhancing ant attraction and thus increasing the effectiveness of this dispersal method. Such a set of adaptations is called a *myrmecochorous syndrome* (Gorb and Gorb, 2001b).

3. MECHANISMS OF ANT ATTRACTION

The main component of the myrmecochorous syndrome is the presence of specialised fat bodies (elaiosomes) on the diaspores (Fig. 1.2) (Sernander,

1906; Ulbrich, 1928; Thompson, 1981; Howe and Smallwood, 1982). The elaiosome is a combined term, covering morphologically-variable structures of plant seeds and fruits. These structures are different in their origin, but both consist of parenchymatous lipid-rich cells (Komar, 1978). Elaiosomes may vary in their shape, coloration, size etc. (Tahtadzhjan, 1985).

3.1 Diversity of elaiosomes

Most myrmecochorous diaspores bear just one elaiosome. However, diaspores of *Polygala vulgaris*, *Triodia decumbens,* and *Stemona tuberosa* are equipped with two appendages (Ridley, 1930; Huber, 1969; Tahtadzhjan, 1985). In *Melica nutans* and representatives of the genus *Fedia,* different flowers of the same inflorescence produce various types of diaspores: with or without appendages (Sernander, 1906; Levina, 1957; Roth, 1977). In contrast to the strikingly-coloured appendages of ornithochorous diaspores, elaiosomes of myrmecochorous diaspores are usually whitish or yellowish; they are coloured differently than diaspore bodies, which are mostly dark. In some myrmecochores, such as *Corydalis cava, C. solida, Lamium maculatum,* and *Pulmonaria obscura*, the light colour of elaiosomes contrasts with the dark colour of diaspore bodies. Some species from the genera *Bocconia, Calathea, Chelidonium, Euphorbia, Hibbertia, Mercurialis,* and *Sterculia* possess yellow-, orange-, and red-coloured elaiosomes (Crété, 1944; Nordhagen, 1959; Holttum, 1951; Kapil, 1956a; b; Rao and Schamanna, 1963; Bullock, 1974; Kennedy, 1978; Tahtadzhjan, 1981; 1985; 1988; 1992).

Shape and size of elaiosomes varies in different plant species (Fig. 1.2) and even within the same species (Gorb et al., 1996). For example, in *Pulmonaria obscura,* elaiosome length varies to a large extent, whereas the length of the erem (a part of the disintegrating fruit coenobium) is rather stable (Fig. 1.3). The shape of elaiosomes may be spherical (*P. obscura*), cylindrical (*Scilla sibirica, Knautia arvensis*), ring-like (*Omphalodes scorpioides*), cone-like (*Viola hirta, V. odorata, Ficaria verna, Strophiostoma sparsiflora*), comb-like (*Gagea lutea*) or even worm-like (*Corydalis cava, C. solida*) (Fig. 1.2). Additionally, elaiosomes may be grouped into three further different structural types: laciniate (*Brachychilum* spp.), divided (species from genera *Buxus* and *Dicentra*), and fimbriated (*Croomia* spp.) (Netolitzky, 1926; Martin, 1946; Nordhagen, 1959; Komar, 1965; Berg, 1966; 1969; Huber, 1969; Corner, 1976; Tahtadzhjan, 1982; 1985; 1988; 1991).

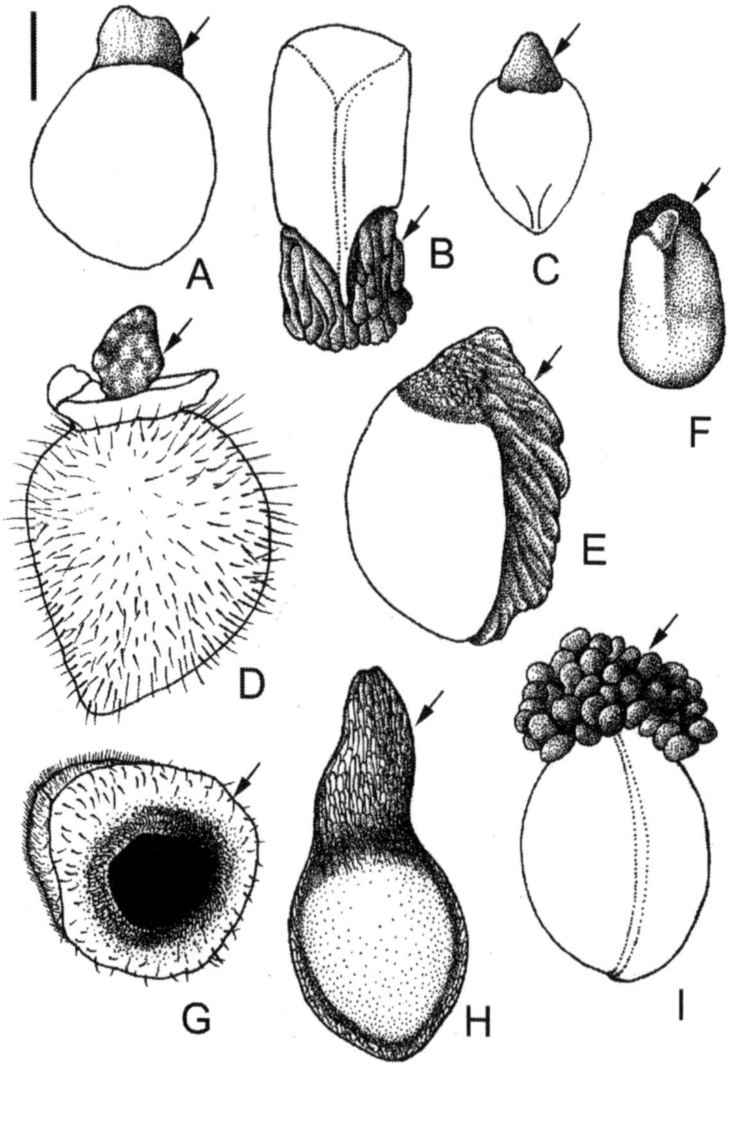

Fig. 1.2. Diaspores of myrmecochorous plants. A. *Scilla sibirica*. B. *Lamium maculatum*. C. *Strofiostoma sparsiflora*. D. *Pulmonaria obscura*. E. *Gagea lutea*. F. *Ballota nigra*. G. *Omphalodes scorpioides*. H. *Ficaria verna*. I. *Scilla bifolia*. Arrows indicate elaiosomes. Scale bar: 1 mm.

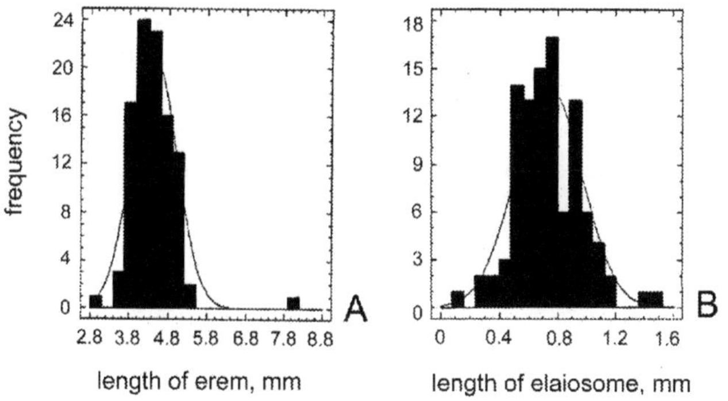

Fig. 1.3. Frequency histograms of erem length (A) and elaiosome length (B) in *Pulmonaria obscura*. Lines are curves of normal distribution.

Origins of fat bodies also vary. In some plants they may appear from various parts of the ovule, pericarp, and even flower and inflorescence (Roth, 1977; Tahtadzhjan, 1985; 1988; 1991; 1992; Werker, 1997). Diversity of elaiosome origins is a good example of convergent evolution that demonstrates how structures of initially different shape and function may be transformed into similar shapes for similar functions (Handel and Beattie, 1990). Some plant tissues adapted for protection, transport and other functions have been transformed in the course of evolution into lipid storage thereby functioning as bait for ant workers.

3.2 Elaiosomes originating from seed tissues

Previous authors used different terminology for elaiosomes of different origin. We follow the terminology of the comprehensive monograph Comparative anatomy of seeds (Tahtadzhjan, 1985). If a seed serves as a functional diaspore, the term «aril» for elaiosome is used. *Aril* or *arillus* is a structure occurring in seeds of many flowering plants. It consists of sappy and/or pulpy tissues or is present as a kind of a pellicle or a fringe. Arils have evolved in various seed parts, and, depending on their origin, are called *true arils* (or just *aril*) or *false arils* (or *pseudoaril*) (Komar, 1965; Tahtadzhjan, 1985) (Fig. 1.4). Additionally, there is a complex type, combining *caruncle* and true aril in the region of the hilum. Arils having dual origin are often termed *complex arils* (Kapil et al., 1980). Complex arils, originating from both the funicle and placenta, have been described in seeds of myrmecochores from the genera *Podophyllum* and *Vancouveria*. Those, originating from both the funicle and raphe, are known in the genera *Bocconia, Croomia, Dicentra*, and *Fumaria* (Sernander, 1906; Crété, 1944;

Nordhagen, 1959; Rao and Schamanna, 1963; Komar, 1965; Berg, 1966; 1969; Huber, 1969; Bullock, 1974; Corner, 1976; Tahtadzhjan, 1985; 1988).

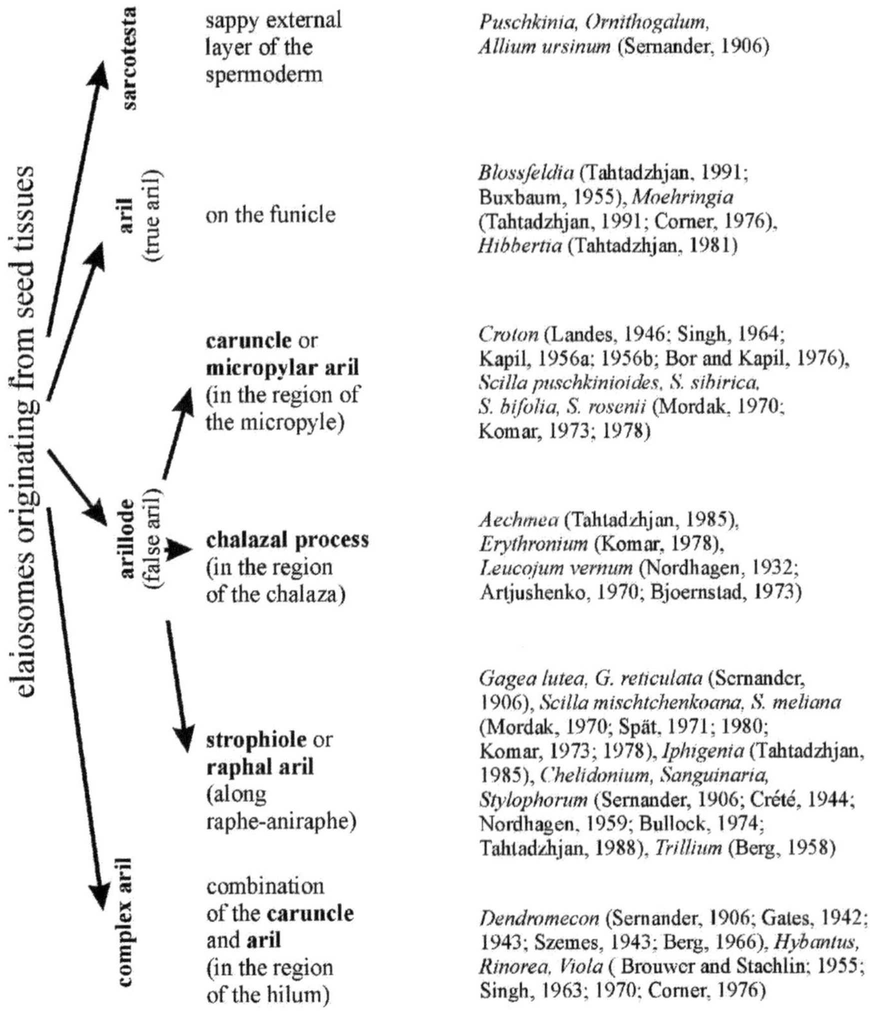

Fig. 1.4. Origins of elaiosomes that are seed appendages.

Arils usually develop after fertilisation. However, they may develop in the form of a thickening prior to fertilisation (Komar, 1965; Werker, 1997). True arils originate either entirely from the epidermal layer or from both external and internal layers. In the case of epidermal origins, the aril develops mostly due to cell elongation and rarely due to cell division (Sernander, 1906; Berg, 1958; 1966; 1969; Bresinsky, 1963). Usually, cells of arils lack nuclei (Appendix, Figs. I, A; II, C), and only a few myrmecochores possess these cells with nuclei (Berg, 1969). In the latter case, cells become polyploid and develop very large nuclei. In *Gagea* sp., elaiosome cells contain 16–32-ploid nuclei. In *Corydalis cava,* the ratio between the dimensions of a diploid nucleus and those of elaiosome cells may reach 1:400 (Geitler, 1944; 1949; 1950).

A typical aril consists of an *epidermis*, a *distinct cortex* (Appendix, Figs. I; II, C), and, in some species, a *medulla* (Werker, 1997). Epidermal cells have various shapes and cell wall thicknesses (Appendix, Fig. I). Cortical cells are normally thin-walled, always pulpy and strongly vacuolated (Appendix, Fig. I, B). The cortex may be further differentiated into sublayers. Some arils may lack transporting tissues, others may be completely traversed by vascular bundles, as in the case of the true aril.

In some myrmecochores, ant-attracting substances are located in the external layers of the seed. These tissues tend to remain soft and sappy after ripening (*sarcotesta*). Sarcotesta-bearing seeds have been described in species from the genera *Cyclamen, Ornithogalum,* and *Puschkinia* (Sernander, 1906; Huber, 1969; Pijl, 1982; Tahtadzhjan, 1985). Elaiosomes of some representatives of the family Euphorbiaceae (*Acalypha, Euphorbia, Mercurialis*) combine sarcotesta and caruncle (Landes, 1946; Kapil, 1956a; b; Tahtadzhjan, 1992) (Appendix, Fig. II, A). The sarcotesta usually evolves from the external layers of the testa and consists of large thin-walled living cells bearing nuclei and vacuoles (Werker, 1997) (Appendix, Fig. II, B). Regions of the spermoderm, underlying the sarcotesta, contain thick-walled sclerenchymatous cells packed with crystals (Corner, 1976).

3.3 Elaiosomes originating from fruit tissues

The literature contains much less information about elaiosomes originating from fruit tissues. There are only a few studies focused on their origin and anatomy (Baciu, 1931; 1932; 1935; Bresinsky, 1963; Roth, 1977). If the functional diaspore is a fruit or fruit part (erem), elaiosomes may originate from various structures of the diaspore (Fig. 1.5). Unfortunately, a good terminology based on the origin of these elaiosomes is still lacking.

Fruit-originating elaiosomes contain four regions with different structures and functions: (1) *peripheral feeding layer*, (2) *fundamental tissue*, (3) *protective tissues*, and (4) *supporting tissue* (Bresinsky, 1963) (Appendix, Fig. III, A). Not all four regions are present in all myrmecochores.

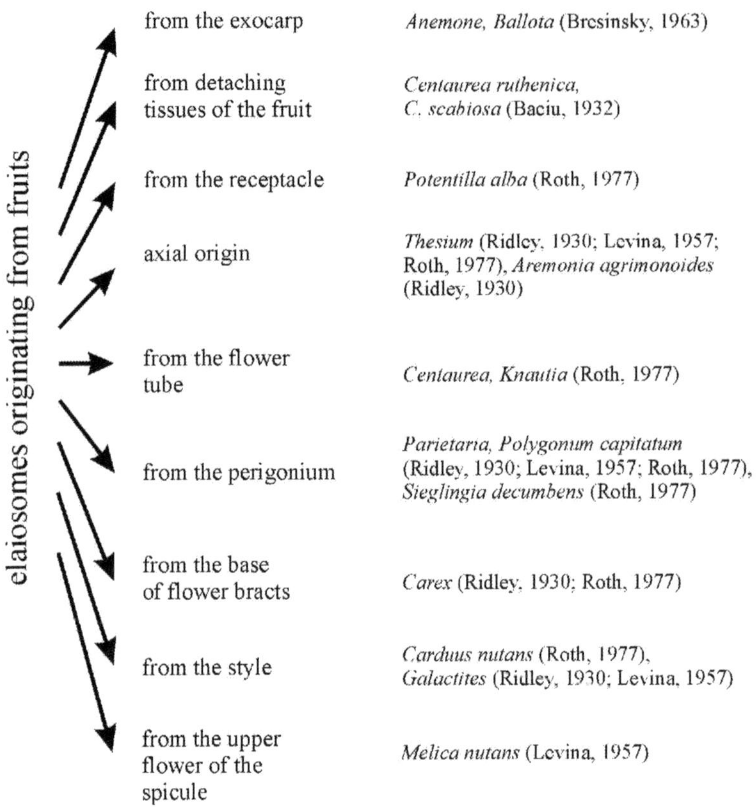

Fig. 1.5. Origins of elaiosomes that are fruit appendages.

The external or peripheral tissue may be one- or multilayered. It usually consists of rather large or even hypertrophied cells with a large amount of fatty substances (*Lamium* spp., *Pulmonaria* spp.) (Appendix, Fig. III, A, B). Sometimes this region is fused with underlying layers and can hardly be distinguished from the latter (Appendix, Fig. IV, B). The fundamental tissue includes thin- or thick-layered cells of various shapes, usually elongated ones (*Anemone hepatica, Centaurea montana, C. cyanus*) (Appendix, Figs. III, C; IV, B, C). The protective tissue is composed of cell layers with heavily-thickened cell walls (Appendix, Figs. III, A; IV, A). This tissue is often called crystal-bearing tissue because the cells contain calcium oxalate crystals providing protection for the endosperm and embryo. The vascular bundle may be involved in the transport of substances only if it runs through the centre of the elaiosome (*Pulmonaria* spp.). In addition, it may provide mechanical stability, if the elaiosome is attached to the fruit through the vascular bundle (*Anemone* spp.) (Appendix, Figs. III, A; IV, C).

3.4 Chemical composition of elaiosomes

Diaspore appendages of myrmecochorous plants have been called elaiosomes (fat bodies) mainly due to the high content of fatty substances in their cells (Sernander, 1906). Fats are present as cell contents or may be connected to the cell wall (Sernander, 1906; Nordhagen, 1932; Bresinsky, 1963) (Appendix, Fig. V, A) or located in spherosomes (Werker, 1997) (Appendix, Figs. V, B; VI). Spherosomes often may reach very large dimensions. In some myrmecochores (*Lamium maculatum, Knautia arvensis, Viola* ssp.), the elaiosome epidermis is covered by a lipid layer of variable thickness (Appendix, Figs. V, B; VII).

Moreover, elaiosome cells bear other reserve substances, such as carbohydrates, proteins, and vitamins (Pijl, 1982). Other studies show that elaiosome cells contain large amounts of fats, fatty acids, and carbohydrates in the form of starch (especially in regions located closely to vascular bundles), and various sugars (glucose, fructose, sucrose) (Bresinsky, 1963). Vitamins B_1 and C are, reportedly, common substances of elaiosome cells. In *Carica papaya* and some representatives of the family Euphorbiaceae, the sarcotesta contains seed germination inhibitors (Lange, 1961; Logoa and Pereira, 1987; Pacini, 1990).

3.5 Elaiosome compounds attracting ants

The main role of lipid substances in ant attraction was hypothesised for the first time by Sernander (1906). Experiments with *Viola odorata* seeds and *Lasius fuliginosus* ants provided evidence of this hypothesis (Bresinsky, 1963). The main conclusion of this study was that ricinoleic acid is an attractant for ant workers. However, experiments with *V. odorata* and ants from the genus *Aphaenogaster* showed that the principal releaser of ant carrying behaviour is not the ricinoleic acid, only a small amount of which is present in elaiosomes, but rather diglycerides especially 1,2-diolein. These substances are responsible not only for ant attraction, but also for releasing carrying behaviour in ant workers (Marschall et al., 1979). In elaiosomes of *Hardenbergia violaceae*, oleic acid makes up 60% of all fatty acids (Hocking and Kortt, 1987). Diolein is the principal component of the diglyceride fraction in elaiosomes of *Hepatica americana,* and an evident chemical releaser of carrying behaviour in ant workers *Pogonomyrmex rugosus* (Skidmore and Heithaus, 1988). Studies that combined chemical analysis of the elaiosome content and observation of the diaspore removal in *Acacia myrtifolia* and *Tetratheca stenocarpa,* have supported the fact that only the fraction of polar lipids, in which 1,2-diglycerides comprise the principal group of fatty acids, stimulates diaspore removal by ants (Brew et al., 1989).

Chemical analysis of lipids in seed bodies and elaiosomes of 12 species of myrmecochorous plants demonstrated that oleic acid is the main portion of the dry weight of seed bodies (38.4%) and elaiosomes (38.9%) (Hughes et al., 1994). Interestingly, fatty acid composition of elaiosomes is very similar to that of the insect body especially in terms of an equal proportion of oleic acid and some other acids (Fig. 1.6). Different species of ants demonstrated a different response to 1,2-diolein: the response was stronger in species which are active seed-collectors, while nectar-feeding species had a much lower response. Such a different response has supported results of previous authors about the stimulating effect of this substance on seed-collecting ants.

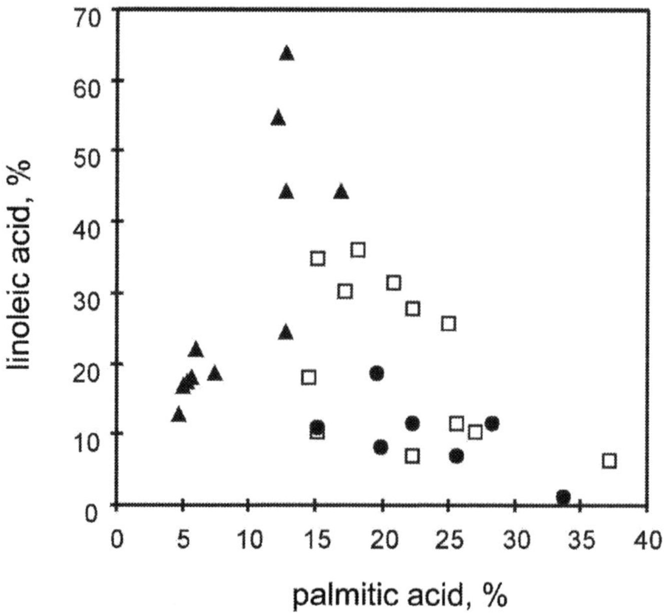

Fig. 1.6. Relationship between content of palmitic and linoleic acids for seeds (12 plant species), elaiosomes (12 plant species) and insects (7 species). Points, insects; squares, elaiosomes; triangles, seeds (data from Hughes et al., 1994).

It is well known that fatty acids are an obligatory component of insect diet (Dadd, 1973) and that some of them are food attractants for ants (Vinson et al., 1967). Since diglycerides are a principal class of neutral lipids in the insect haemolymph (Gilbert and Chino, 1974), and oleic acid is the principal component of fatty acids of the haemolymph (Thompson, 1973), ants presumably respond to diglycerides as a food source (Marschall et al., 1979; Hughes et al., 1994). Elaiosomes not only chemically imitate a victim for predatory ants: their high fat content provides energetic value of the elaiosomes as a potential food object. Since lipids are known to be attractants and pheromones (Dethier, 1947; Gilbert, 1967), substances, found in elaiosomes, may release specific ant responses, such as brood transport into the nest (Carthy, 1951). Because the strongest ant response is to 1,2-diolein and not to the entire spectrum of diglycerides of the elaiosome, it is obvious that this particular substance is a key chemical stimulant for ants (Marschall et al., 1979).

3.6 Effect of diaspore and elaiosome dimensions on diaspore attractiveness for ants

Diaspores of different myrmecochorous plant species have different dimensions of diaspore bodies, elaiosomes, and different ratios between these dimensions (Fig. 1.7). Some authors showed that elaiosome size (Davidson and Morton, 1984; Gunther and Lanza, 1989; Oostermeijer, 1989; Mark and Olesen, 1996) and the ratio of elaiosome size to the diaspore body size (Hughes and Westoby, 1992) may influence diaspore removal rate by ants. Our data demonstrated the same effect for ant-plant relationships of the temperate deciduous forest (Gorb and Gorb, 1995a; b; Gorb, 1998). However, experiments with the removal by ants of *Viola hirta* seeds with partly ablated elaiosomes or diaspore bodies revealed that seed attractiveness mostly depends on the presence, but not on the size, of the elaiosome. Elaiosome size appears to be mainly important at the stage of seed removal (Gorb and Gorb, 1995b).

Diaspores, adapted to dispersal by ants, usually belong to a very narrow band of the spectrum of size classes (Table 1.1). This may be explained by the fact that diaspores have to be large enough to be spotted by ants, and at the same time, light enough to be removed by ant workers. Diaspore dimensions play a crucial role for ant attraction during the searching activity of foragers. Curvature of the diaspore is the most important variable in the stage of the removal and transport (Gorb and Gorb, 1995b). Experimental results on the seed removal of several myrmecochorous plant species by *Formica polyctena* ants revealed that large seeds (*Asarum europaeum, Corydalis cava, Viola hirta*) were more often taken by ants than small ones (*C. solida, Chelidonium majus, V. mirabilis, V. matutina*) (Gorb and Gorb, 1995a; b; Gorb, 1998).

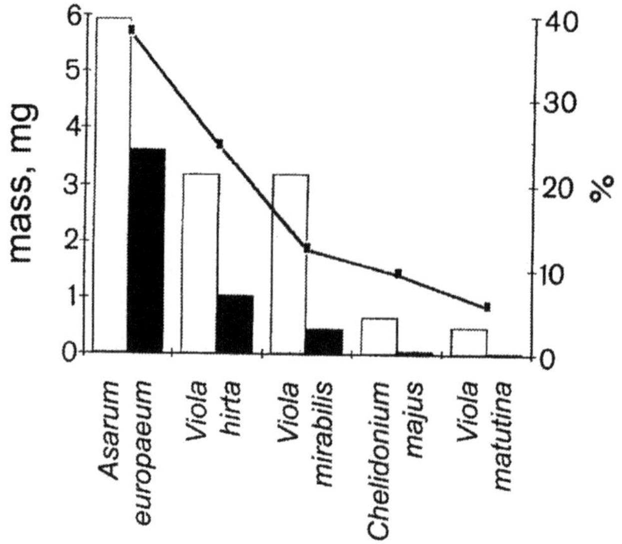

Fig 1.7. Relationship between total seed mass and elaiosome mass in five plant species. White columns, total seed mass; black columns, elaiosome mass; line, percentage of elaiosome mass in relation to seed mass.

The fact that the presence of elaiosomes often correlates with other plant adaptations to myrmecochory was stated by Ulbrich (1928). Among such adaptations the most conspicuous ones are (1) drooping and lowering of pedicles, (2) early periods of fruiting, (3) development of additional assimilating organs. These adaptations are considered below.

4. DIASPORE DEPOSITION INTO MICROSITES FREQUENTLY VISITED BY ANTS

Effectiveness of seed removal is enhanced by their location at those microsites most frequently visited by ants. Pedicles of many species of myrmecochores have underdeveloped mechanical tissues. Because of this, at the moment of dissemination, pedicles do not stiffen and/or do not elongate, but simply bend down to the soil under the weight of fruits. This occurs when turgor pressure subsides in the fleshy tissues of the pedicles (Ulbrich, 1928; Levina, 1957; Handel and Beattie, 1990).

Pedicles of myrmecochores usually bend down (*carpotropic bends*), and the fruits lie on the soil or the litter surface (Fig. 1.8). In some species (*Carex umbellata, Trillium petiolatum*) the pedicles are so short that diaspores ripen just above the ground (Handel and Beattie, 1990). In Australian myrmecochorous acacias, seeds lack special adaptations for displaying them on the plant (characteristic for ornithochorous species of acacias), and seeds are deposited on the soil surface, where they are most accessible to ants (Davidson and Morton, 1984).

Table 1.1. Diaspore dimensions in 18 species of myrmecochorous plants. Abbreviations: n, sample size; sd, standard deviation; x_{ave}, average values.

Plant species	Diaspore length, mm			Diaspore width, mm		
	x_{ave}	sd	n	x_{ave}	sd	n
Knautia arvensis	5.08	0.59	50	2.13	0.19	50
Pulmonaria obscura	4.44	0.60	100	3.67	0.40	100
Gagea lutea	3.83	4.23	50	1.74	0.29	50
Asarum europaeum	3.77	0.24	80	2.64	0.28	80
Lamium maculatum	3.45	0.22	30	1.70	0.11	30
Mercurialis perennis	3.19	0.31	21	3.37	0.20	21
Stellaria holostea	2.82	0.23	50	2.11	0.19	50
Corydalis cava	2.75	0.21	96	3.17	0.24	96
Scilla bifolia	2.73	0.22	35	2.40	0.19	35
S. sibirica	2.67	0.21	50	2.27	0.23	50
Viola odorata	2.47	0.18	30	1.99	0.16	30
Ficaria verna	2.36	0.25	25	2.06	1.22	25
V. hirta	2.00	0.10	30	1.38	0.10	30
Strophiostoma sparsiflora	1.97	0.18	30	1.37	0.22	30
Corydalis bulbosa	1.72	0.14	90	1.71	0.14	90
Chelidonium majus	1.59	0.05	30	1.06	0.05	30
Omphalodes scorpioides	1.32	0.15	28	2.37	0.18	28
V. matutina	1.31	0.11	50	0.81	0.08	48

Since different ant species have different foraging niches (Dlussky, 1967), seed deposition by plants may strongly influence ant visits. We performed field experiments (for a detailed method description see chapter 11, section 4.2.1), where ant visiting rates of seed depots of the violet *Viola odorata* were compared between depots located in the litter with those on the litter of a deciduous forest. Interestingly, different depots were visited at various rates by ants of different species (Gorb and Gorb, 1998a; Gorb, 2000) (Fig. 1.9). Thus, diaspore deposition by the plant either on or in the litter may predetermine potential seed dispersing ants.

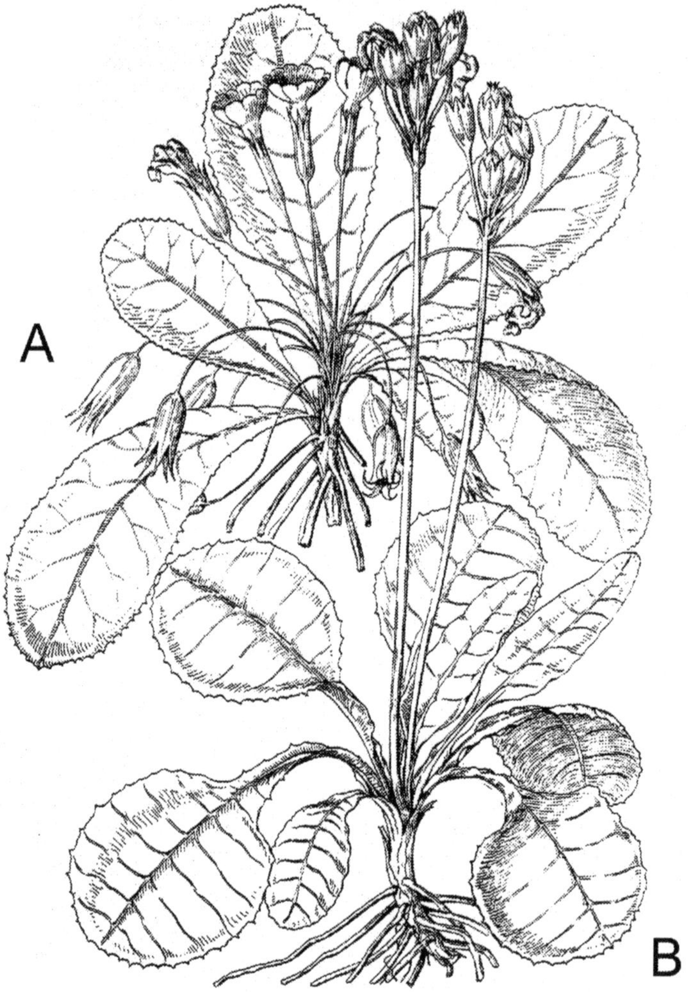

Fig. 1.8. Two herbs with different diaspore dispersal mechanisms: myrmecochore *Primula acaulis* (A) and non-myrmecochore *P. elatior* (B). In the myrmecochorous species, pedicles are turned down and flower bells are considerably larger (from Ulbrich, 1928). With permission of Springer Verlag.

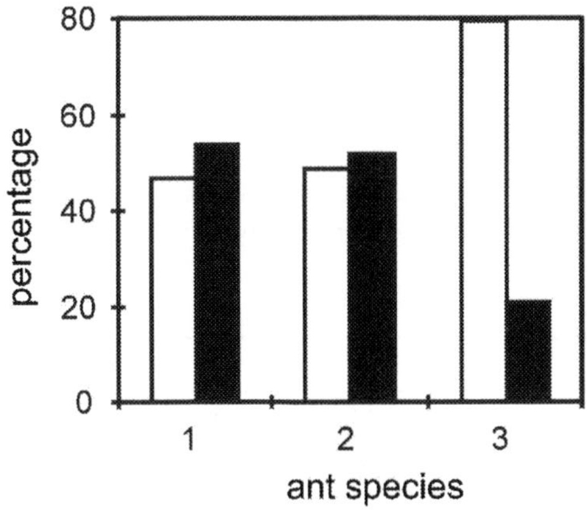

Fig. 1.9. Relationship between the number of ants visiting seed sets located in (black columns) and on (white columns) forest litter. 1, *Myrmica rubra*; 2, *Leptothorax nylanderi*; 3, *Formica polyctena*.

5. SYNCRONISATION OF THE PLANT FRUITING PERIODS WITH PERIODS OF ANT ACTIVITY

Many myrmecochorous plants have early flowering and fruiting periods. In temperate deciduous forests, diaspores of most ant-dispersing plants ripen in the late Spring (Ulbrich, 1928; Handel and Beattie, 1990; Gorb, 1998) (Fig. 1.10). In meadows and steppes, myrmecochores disperse their diaspores in the late Spring to early Summer (Levina, 1957). Diaspore dispersal by ants is a wide-spread phenomenon especially among ephemeral spring plants (Thompson, 1981). A shift in the fruiting period in myrmecochores to earlier periods may be explained by higher activity of ant foragers at these periods of time. A comparison of the dynamics of the brood emergence and the fat content in workers with the fruiting periods of myrmecochores clearly shows that the dissemination period of myrmecochores fits exactly to the most critical period in the colony resources (Kirchner, 1964) (Figs. 1.10; 1.11).

In the Spring, workers of the ant *Formica polyctena* after wintering begin to breed sexual individuals by using resources accumulated during the past Autumn (Dlussky, 1967). Immediately after the emergence of winged individuals, the ant colony proceeds with the mass breeding of workers. This period extends from Spring to the first half of Summer. During this period, ants mainly collect food for larval individuals. Fat content in the workers' bodies drastically drops (Fig. 1.11). Aphid colonies are still relatively small at this time, and seed elaiosomes fill the gap in the nutrition of workers (Oekland, 1932; Dlussky, 1967). Because the number of insects, which are normally the main component of ant diet, is also relatively low at this time compared to the middle of Summer, elaiosomes are actively used as the food for larval individuals. This may be the reason for active transporting of seeds to ant nests

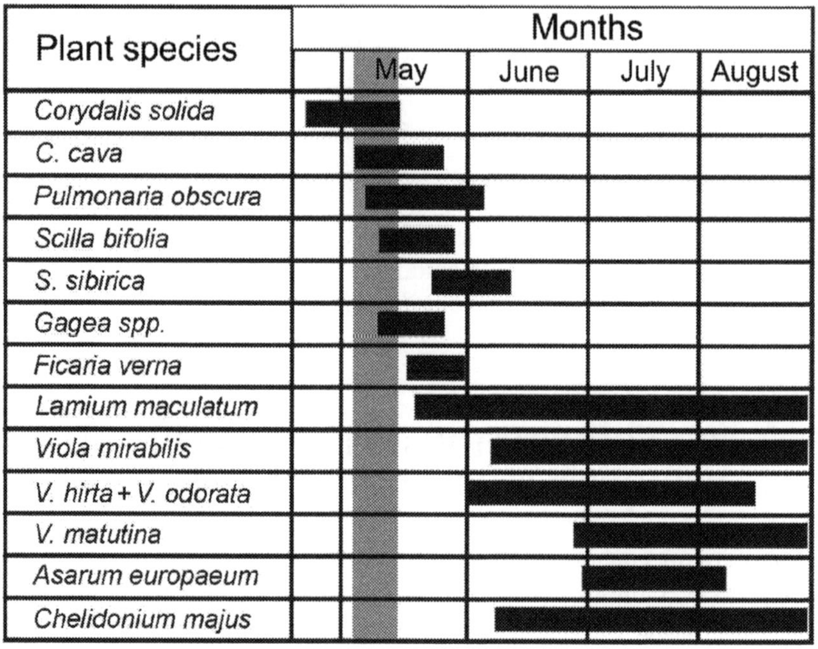

Fig. 1.10. Phenology of fruiting of some herbaceous myrmecochores of the deciduous forest.

Data on seed dispersal in *Corydalis cava* and *C. bulbosa* show that plants, whose diaspores ripen in early Spring, may escape high competition with other myrmecochorous plants for dispersal agents (Gorb and Gorb, 1995a; Gorb, 1998). The dissemination period of *C. ambiqua* (late May) correlates well with the period of high ant density and the lowest density of seed-damaging ground beetles (Ohkawara et al., 1997). In *Viola nuttallii,* the time of day of fruit opening is synchronised with the time of the highest activity of ants and the lowest seed-eating and non-dispersal activity of rodents. This mechanism provides the most rapid seed removal by dispersers and prevents seed damage by other animals (Turnbull and Culver, 1983).

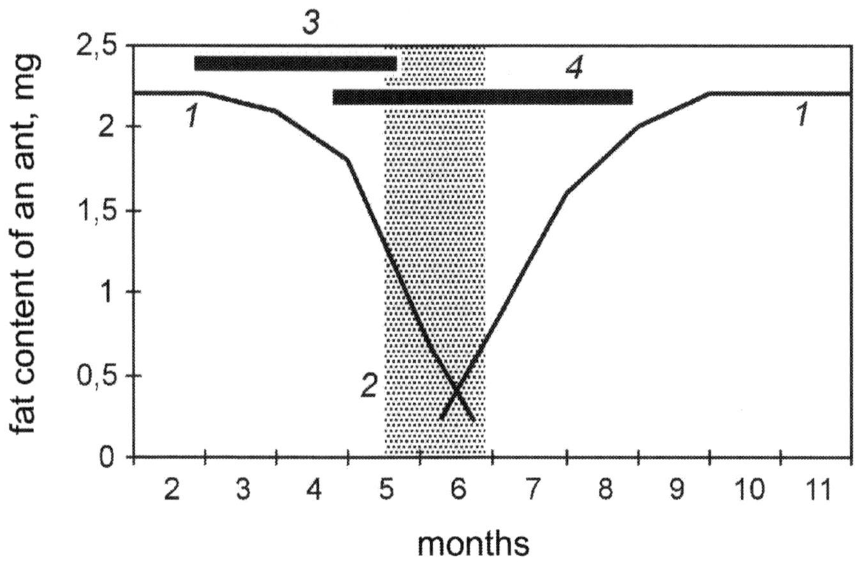

Fig. 1.11. Relationship between ant colony behaviour and phenology of fruiting of myrmecochorous plants. 1, seasonal changes in the fat content in *Formica polyctena* workers; 2, peak of the myrmecochores fruiting in temperate forests; 3, period of time when sexual individuals are fed. 4, time of feeding of workers. 1, 3, 4, data from Dlussky (1967).

6. SUPPLEMENTARY ASSIMILATING ORGANS

In the majority of plant species, diaspore pericarps dry out and serve only mechanical function as a seed container at the moment of dissemination. Since the production of large diaspores with large elaiosomes requires considerable inputs of organic substances, in some myrmecochores, such as *Iris aphylla* and *Viola odorata,* capsules and other fruit types remain green and photosynthetic until seeds are completely ripe (Levina, 1957). In some other species, flower bell (*Borago* spp., *Anemone nemorosa*) or apical leaves (*Hepatica* spp.) are largely increased after flowering and transformed into additional assimilating organs (Ulbrich, 1928; Levina, 1957). Such adaptations occurred most frequently among myrmecochorous plants.

7. THE MYRMECOCHOROUS SYNDROME IN FACULTATIVE MYRMECOCHORES

Plant adaptations for diaspore dispersal by ants are best expressed among obligate myrmecochores exploiting only this method of dispersal (Fig. 1.1). Diaspores of various plant species have a different degree of attractiveness for one particular ant species and for the entire ant community (Gorb and Gorb, 1995a; b; Gorb, 1998). This may be explained by unequal expression of myrmecochorous syndrome characters in plants. Furthermore, in some plant species, myrmecochory is supplemented by other methods of diaspore dispersal (*facultative myrmecochory*, *diplochory*), such as *anemochory* (Ulbrich, 1928; Levina, 1957), *endozoochory* by birds (Davidson and Morton, 1984) and primates (Roberts and Heithaus, 1986). Most frequently myrmecochory is supplemented by autochory (Sernander, 1906; Ulbrich, 1928; Levina, 1957; Berg, 1966; Beattie and Lyons, 1975; Beattie and Culver, 1979; Westoby and Rice, 1981; Oostermeijer, 1989; Ohkawara and Higashi, 1994; Passos and Ferreira, 1996).

In facultative myrmecochory, some characters of the myrmecochorous syndrome may be less expressed compared with other adaptations for other methods of diaspore dispersal or even completely lost. For example, diplochorous violets bear very small seeds with minute elaiosomes. Their capsules contain well-developed mechanical tissues; flower bells do not increase in size after flowering; pedicles are erect, long, often thickened (Beattie and Lyons, 1975; Gorb and Gorb, 1995b). Another example is a diplochorous species from the genus *Dendromecon* having little seeds with rather large elaiosomes. The elaiosomes, however, drastically decrease in size prior to the dissemination. Capsule walls, after fruit ripening, become thickened and dry (Berg, 1966). In *Melampyrum argyrocomum*, diaspores bear well-developed elaiosomes. However, pedicles during flowering and ripening periods remain erect, and the fruits face upwards (Levina, 1957). Such an orientation of fruits complicates seed removal by ants. A peculiar compromise between adaptations to different dispersal methods usually leads to a decrease of diaspore attraction for ants and to lower removal rates of diaspores of facultative myrmecochores by ants (Gorb and Gorb, 1995b).

8. OTHER EXAMPLES OF ANT-PLANT RELATIONSHIPS

Few other examples of mutualistic interactions between ants and plants (*myrmecophily*) are known. Some plants attract ants by means of extrafloral nectaries, which serve as food sources for ants. Additionally, plants may provide refuges for ants in expanded parts (*domatia*) of leaves, stems, and bases of thorns; in return ants provide plant protection against herbivores. Typical examples of such a mutualism are representatives from the plant genera *Cecropia* (Davidson et al., 1991; Davidson and Fisher, 1991), *Myrmecodia, Hydnophytum* (Huxley, 1978), *Macaranga* (Federle et al., 1998), and some species of the genera *Piper* (Letourneau, 1998) and *Acacia* (Wagner, 1997). Many myrmecophylic plants, such as *Croton* (Ganeshaiah and Shaanker, 1988), *Besleria* (Windsor and Jolivet, 1996), and *Acacia* (Davidson and Morton, 1984; Andersen, 1988), have an additional advantage from diaspore dispersal by ants. In these cases, typical characters of the myrmecochorous syndrome may be set against a background of other adaptations to the myrmecophily (O'Dowd and Gill, 1986).

Another type of myrmecophilic interaction is well known under the term *ant gardens*. It was discovered in a series of tropical epiphytic plants which are closely-connected in their biology to arboreal ants (Ulbrich, 1928; Kleinfeldt, 1986; Davidson, 1988). In this case, ants use soil particles and seeds of a particular plant species for building their nests in the canopy on tree branches or even on gigantic palm leaves. Seeds of these plants possess typical characters of the myrmecochorous syndrome described above (Davidson et al., 1990; Seidel et al., 1990). Roots of plants, germinated from these seeds, serve to stabilise the nest constructions of ants. Plants use benefits from the aeration of the soil by ants and from the high nutrition of the nest soil containing organic wastes of the ant colony (Yu, 1994).

9. THE MYRMECOCHOROUS SYNDROME AND CRITERIA OF MYRMECOCHORY

According to the fundamental criteria of myrmecophily, which may be applied equally to myrmecochory: plants have to (1) take evident advantages of the ants' activity, and (2) possess specialised structures for attracting ants (Blatter, 1928). According to the first criterion, there are five hypotheses about *selective advantages of myrmecochory* (those promoted by natural selection): (a) *occupation by plants of nutrient-rich microsites on ant nests* (the nutrient-rich nest environment hypothesis); (b) *avoidance of diaspore consumption by predator animals* (the predator-avoidance hypothesis); (c) *avoiding competition between seedlings* (the competition-avoidance hypothesis); (d) *avoiding fire* (the fire-avoidance hypothesis); (e) *dispersal for distance from the parent plant* (dispersal for distance hypothesis) (Beattie, 1983; 1985; Handel and Beattie, 1990). Over the past years, different authors provided experimental evidence supporting or rejecting some of the named hypotheses (see chapter 9, section 1.2). In spite of the numerous studies devoted to plant adaptations to myrmecochory, the exact significance of single characters of the myrmecochorous syndrome for particular selective advantages remains poorly studied.

According to the second criterion, characters of myrmecochorous syndrome are responsible for ant attraction and for enhancement of seed transport by ants. In numerous studies, effects of particular characters of the syndrome in single plant species on the diaspore dispersal by ants have been described (Davidson and Morton, 1984; Gunther and Lanza, 1989; Oostermeijer, 1989; Hughes and Westoby, 1992; Mark and Olesen, 1996). In the following chapters we summarise the results of laboratory studies and field experiments in order to evaluate the specific role of each character within the total range of putative adaptations for diaspore dispersal by one particular ant species and ant community. Since the experiments were carried out in the same ecosystem, the data from different experiments supplement one another thus reaching the most complete picture of the phenomenon of myrmecochory in the deciduous forest.

10. SUMMARY

- Myrmecochorous plants show morphological, anatomical, biochemical and phenological features connected to ant dispersal. The complex of plant adaptations to this dispersal method is called myrmecochorous syndrome. This chapter summarises the diversity of plant adaptations and analyses the role of these adaptations for the attraction of ants.
- The principal feature of the myrmecochorous syndrome is an elaiosome (a fat body) attached to the diaspore. Different myrmecochores have elaiosomes of diverse shapes, colours, sizes, anatomy, and origin. Different parts of ovule, pericarp or even flower and inflorescence may be initial tissues for the elaiosome origin.
- Elaiosomes contain a particular complex of fatty acids (mainly 1,2-diolein) which function as triggers for the carrying behaviour of ant workers.
- Seed setting periods, which are synchronised with periods of maximal ant activity, and particular sizes of diaspores aid in ant attraction and in localisation of diaspores by ants.
- In the case of a facultative myrmecochory, when a plant has additional mechanisms of diaspore dispersal, some characters of the rmecochorous syndrome may be absent or less expressed.

Chapter 2

Diaspore removal by ants

The first stage of the myrmecochory is the removal of diaspores from the fruit or from the forest litter by ant workers. This stage can be quantitatively described by the *removal rate*, which is expressed as the number of diaspores taken by ants per unit of time. Previous authors have focused their experiments mainly on the role of environmental factors, such as weather, season, or time of day on removal rate (Hughes and Westoby, 1990), and on the interaction between a given plant species and a particular ant species (Kjellsson, 1985; Pacini, 1990; Mark and Olesen, 1996) or between one plant species and a complex of ant species (Culver and Beattie, 1980; Nesom, 1981; Clay, 1983; Bond and Stock, 1989; Higashi et al., 1989; Wein and Pickett, 1989). Very few studies have been undertaken to clarify the relationship between a complex of plant species and a single ant species (Gunther and Lanza, 1989; Baroni-Urbani and Nielsen, 1990). The present chapter is based mainly on the results of the authors' field experiments, in which some aspects of interactions between the complex of plant species and the complex of ant species in the eastern European (Central Ukraine) deciduous forest were studied at the stage of seed removal.

1. «CAFETERIA» EXPERIMENTS

Previous authors have mentioned the fact that diaspores of the same plant species may be removed by different ant species. This means that ant-plant interactions at the level of diaspore dispersal may not be species-specific. Different species of myrmecochorous plants vary in the anatomy of diaspores, and representatives of different ant taxa also vary in their morphology and biology. Such variability of plants and their potential dispersal agents make interactions between plant complexes and ant communities of the deciduous forest quite complicated. It can be hypothesised that, depending on the ant species complex of forest microsites, the ant community influences the distribution of herbaceous plant species through diaspore dispersal.

To prove this hypothesis, «cafeteria» experiments were performed in the field (for a detailed method description see chapter 11, section 4.2.2). In this type of experiment, food objects are presented to the ants, and the removal events of single objects are recorded over a period of time. This type of experiment is widely used in studies of food preferences and foraging behaviour of ants (Culver and Beattie, 1978; Ashton, 1979; Drake, 1981; Andersen, 1982; Kjellsson, 1985; Hughes and Westoby, 1990). In our experimental set-up, four principal comparisons were made:

(1) Comparison of diaspore removal rates for five plant species (*Asarum europaeum, Chelidonium majus, Viola matutina, V. mirabilis, V. hirta*) by workers of *Formica polyctena* ants. Plant species selected had different seed sizes and anatomy as well as elaiosome origin and design. This comparison was based on solitary and paired presentations of diaspores of different species.

(2) Comparison of diaspore removal rates at some distance from *F. polyctena* nests (*F. polyctena* mainly removed seeds, however, many other ant species were able to remove diaspores) and in the vicinity of nests (only *F. polyctena* removed seeds).

(3) Comparison of the removal rates of intact seeds of *V. hirta* and seeds with partly ablated elaiosome or seeds with partly ablated seed bodies.

(4) Comparison of the removal rates of diaspores of different plant species by *F. polyctena* ant workers of different sizes.

2. MORPHOLOGY OF DIASPORES OF MYRMECOCHOROUS PLANTS

Plant species, the diaspores of which were used in the experiments, belong to two principal groups according to the dispersal strategies employed: obligate (*A. europaeum* and *V. hirta*) and facultative myrmecochores (*Ch. majus, V. matutina,* and *V. mirabilis*). Obligate myrmecochores employ diaspore dispersal by ants as a single dispersal strategy, whereas facultative myrmecochores may use additional dispersal methods. In obligate myrmecochores, pedicles may lie down and/or droop (see also chapter 1, section 4). However, diaspores of some obligate myrmecochores have fruits that ripen in the soil (*geocarpic* plants) or fall off by gravity (*autobarochory*) (terms after Levina, 1957; 1967). Facultative myrmecochores employ a range of dispersal methods, including zoochory, autochory (*automechanochory*), anemochory, *hydrochory*, etc., in addition to myrmecochory.

Obligate myrmecochores usually have large diaspores with relatively large elaiosomes (Table 1.1). Among species used in the «cafeteria» experiments, the seed of *A. europaeum* is oval-triangular, slightly flattened and dark-brownish, with a medially-attached elaiosome (Fig. 2.1, B). In *V. hirta,* the seed is ovate and yellow-brownish, with a basally-attached elaiosome (Fig. 2.1, A).

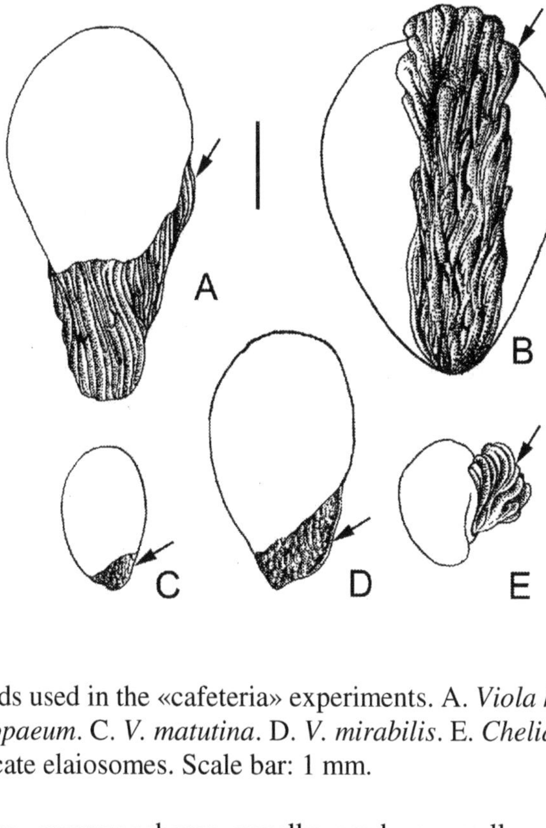

Fig. 2.1. Seeds used in the «cafeteria» experiments. A. *Viola hirta*. B. *Asarum europaeum*. C. *V. matutina*. D. *V. mirabilis*. E. *Chelidonium majus*. Arrows indicate elaiosomes. Scale bar: 1 mm.

Facultative myrmecochores usually produce small or medium-sized diaspores with minute elaiosomes (Table 1.1). The seed of *Ch. majus* is oval and shiny-black, with a medially-attached elaiosome (Fig. 2.1, E). In both *V. matutina* and *V. mirabilis,* seeds are oval, brownish or yellow-brownish with basally-attached elaiosomes (Fig. 2.1, C, D). Interestingly, the elaiosome-to-seed-body mass ratio correlates directly with a decrease in the mass of the diaspore (see Fig. 1.7, chapter 1).c

3. REMOVAL RATES OF ELAIOSOME-BEARING SEEDS BY *FORMICA POLYCTENA* WORKERS

In experiments with elaiosome-bearing seeds of *Acacia linifolia* and *Bossiaea obcordata* and an ant community consisting of 16 species, it has been demonstrated that removal rates are almost independent of the plant species (Hughes and Westoby, 1990). Other authors, however, showed that harvesting ants *Messor capitatus* (Baroni-Urbani and Nielsen, 1990), *M. barbatus, M. sanctus* (Cerdan and Delye, 1990), *Pogonomyrmex badius* (Nicke and Neal, 1972), *M. denticornis, Tetramorium rufescens* (Marsh, 1989), and *Solenopsis geminata* (Caroll and Risch, 1984) display a clear preference for particular plant species. Harvesting ants are seed-eating species and collect seeds not necessarily equipped with an elaiosome. This fact is important to consider the presence of the same effects in the forest plant community. Our experiments with *F. polyctena* and five plant species revealed drastic differences in diaspore removal depending on the plant species and diaspore size.

Dynamics of the diaspore removal of plants studied may be described by an exponential function (Figs. 2.2, A – E; 2.3, A). In the series of species *A. europaeum – V. hirta – V. mirabilis – Ch. majus – V. matutina,* the character of the diaspore removal curves tends to change from an exponential to a linear one. The highest removal rates were observed in obligate large-seeded myrmecochores *A. europaeum* and *V. hirta*, the lowest one – in the small-seeded facultative myrmecochore *V. matutina*. To compare removal rates in different plant species, *time of half removal* (THR) was used. THR is the time, in which half the diaspores are removed from the complete diaspore depot (in our case, it was five out of 10 diaspores). The THR reflects well the differences in removal rates. According to the THR, two plant groups can be easily distinguished: large-seeded obligate myrmecochores (*A. europaeum* + *V. hirta*) and small-seeded facultative myrmecochores (*Ch. majus* + *V. matutina*) (Table 2.1). Differences within each group were negligible.

4. REMOVAL RATES OF SEEDS BY THE ANT SPECIES COMPLEX OF THE FOREST

In spite of the fact that *F. polyctena* workers most frequently removed diaspores at the study site, data on diaspore removal dynamics obtained for this species and for the entire ant community of the site were varied. Comparison of data on removal dynamics, obtained in cafeteria experiments at different distances from the *F. polyctena* nest (5 and 15 m), revealed a similarity in the exponential character of removal curves (Figs. 2.2; 2.3). However, at a greater distance from the nest, removal rates were at least 10 times lower than in the vicinity of the nest. This could also be explained by the fact that, that at a greater distance from the nest, the density of *F. polyctena* workers is considerably lower, which is partly compensated for by the presence of smaller ant species (representatives of the genus *Myrmica*).

At both distances from the nest, the highest values of removal rates were found for *A. europaeum* and *V. matutina*. However, diaspore removal for *V. matutina, Ch. majus,* and in particular for *V. mirabilis* was relatively higher at a greater distance from the nest, if compared to other plant species used in the experiments (Figs. 2.2; 2.3). Presumably, in nature, diaspores that show low removal rates in experiments with *F. polyctena* (*V. mirabilis, Ch. majus, V. matutina*), are usually removed by other ant species. Small diaspores of facultative myrmecochores may be actively removed by individuals of smaller ant species, such as *Myrmica rubra* and *Leptothorax nylanderi*. Nevertheless, individuals of very small ants *L. nylanderi* often gnaw away elaiosomes of large-seeded myrmecochores without removing them. This species does not contribute to the seed dispersal of any plant species studied. Furthermore, diaspores with damaged elaiosomes become less attractive to potential dispersers. Therefore, in microsites with a high density of elaiosome thieves, small-seeded myrmecochores will have some advantages, because they will be much more quickly removed than large-seeded myrmecochores. Elaiosome consumption without diaspore dispersal has been described for other ant species, such as *Lasius niger, Camponotus obscripes* (Higashi et al., 1989), and *Lasius pallitarsis* (Mesler and Lu, 1983). Thus, presence of the community of ant species in the forest may balance the opportunities for diaspore dispersal of various plant species present at the microsite. This may explain why, in experiments with many ant species, differences in removal rates between plant species have not been revealed (Hughes and Westoby, 1990).

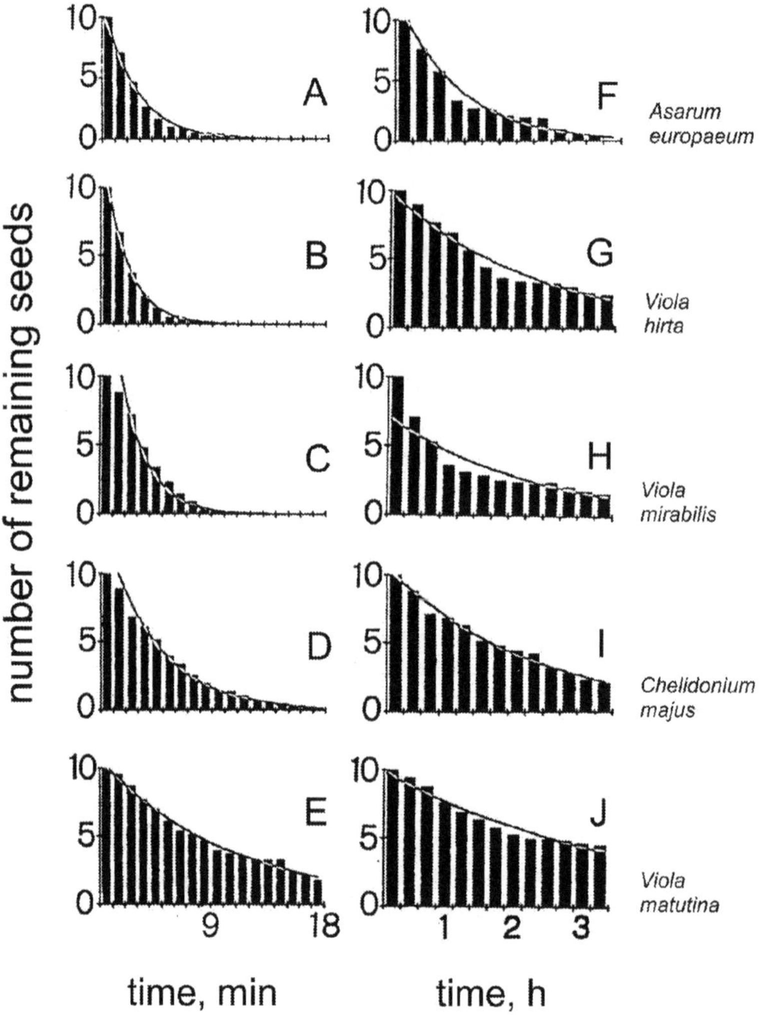

Fig. 2.2. Removal rates of seeds in experiments of solitary presentations for *Formica polyctena* ants close to their nests (A – E) and for the ant species complex of a deciduous forest at an average distance of 15 m from *F. polyctena* nests (F – J). Curves indicate exponential fits calculated for each sample.

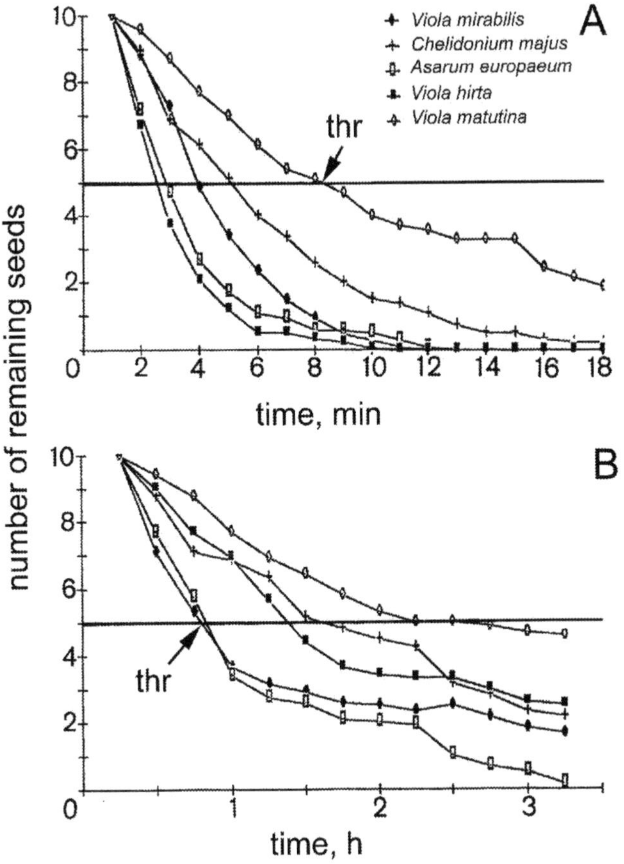

Fig. 2.3. Comparison of the removal rates of seeds in experiments of solitary presentations for only *Formica polyctena* ants (A) and for the entire ant species complex of a deciduous forest (B). thr, the time of half removal (the time of removal of a half of the seed depot). Charts are based on the data presented in Fig. 2.2.

5. DO ANTS PREFER SEEDS OF A PARTICULAR PLANT SPECIES?

It may happen that, in nature, ants simultaneously encounter diaspores of different plant species. Do ants prefer some plant species over others? To clarify this, cafeteria experiments with paired presentations of diaspores from different plant species were performed. Comparison of the values of the time of half removal revealed similar values among the following pairs: *A. europaeum* + *V. hirta*, *A. europaeum* + *V. mirabilis*, *Ch. majus* + *V. mirabilis*, *Ch. majus* + *V. matutina* (Table 2.2). Other pairs have different THR-values. Species with similar curves of the removal dynamics in solitary presentation experiments (Fig. 2.3) usually have similar THR-values in paired presentation experiments. These results led to the conclusion that diaspores with similar removal rates in the solitary presentation experiments, will be removed by ants, in nature, with the same probability.

Table 2.1. Time of the half removal (THR) of diaspores of five plant species in the experiments performed at a distance of 5 m and 15 m from the nest of *Formica polyctena* ant. Abbreviations: n, number of experiments carried out; sd, standard deviation; x_{ave}, average values.

Plant species	Distance from the nest					
	5 m			15 m		
	x_{ave}, s	sd	n	x_{ave}, s	sd	n
Viola hirta	114.1	82.7	22	86.7	50.6	12
Asarum europaeum	137.9	99.5	14	50.8	43.1	12
Viola mirabilis	186.0	48.7	15	46.9	39.8	12
Chelidonium majus	284.3	122.6	14	100.8	64.0	12
Viola matutina	420.0	129.6	7	135.8	50.7	12

Diaspores of plant species with considerable differences in removal dynamics in solitary presentation experiments (*V. hirta* + *V. matutina*, *V. mirabilis* + *V. matutina*, *V. hirta* + *Ch. majus*, *A. europaeum* + *Ch. majus*) revealed even higher differences in the THR-values in the paired presentation experiments (Fig. 2.3; Table 2.2). The lowest THR-value in the paired presentations with all five plant species studied was registered for *V. matutina*. In this case, differences in removal dynamics were most noticeable at an initial stage of the seed depot presentation. This may indicate that, in paired presentations, more attractive diaspores may suppress removal of diaspores of other plant species especially at an initial stage of the

6. DO ANTS SELECT SEEDS WITH THE LARGEST ELAIOSOMES?

Ants much more readily remove larger diaspores with larger elaiosomes (*A. europaeum*, *V. hirta*, *V. mirabilis*). Are they guided by the diaspore size or only by dimensions of the energetically-valuable elaiosome? Previous authors supposed that elaiosome size is the most important cue for ant attraction (Davidson and Morton, 1984; Gunther and Lanza, 1989; Oostermeijer, 1989; Hughes and Westoby, 1992; Mark and Olesen, 1996). Experiments with seeds of *Viola* sp. with completely ablated elaiosomes revealed that ants are not attracted by such seeds (Imhanitskaja, 1981). In our experiments with the paired presentation of intact seeds of *V. hirta* and seeds with partly ablated elaiosomes, no significant difference in removal rates of these two groups of seeds was found. Nevertheless, intact seeds were much more attractive than seeds with partly ablated seed bodies. experiment, when the number of seeds in the depot is relatively high.

Table 2.2. Time of the half removal (THR) of diaspores of five plant species in experiments with pair presentation. Abbreviations: ns, non significant; n, number of experiments carried out; p, criterion of significance; sd, standard deviation; t, value of the Student's criterion; x_{ave}, average values.

Pairs of plant species	x_{ave}, s	sd	n	t	p
Asarum europaeum *Viola hirta*	130 140	40 70	5	0.28	ns
Asarum europaeum *Viola mirabilis*	342 250	116 43	5	1.68	ns
Asarum europaeum *Viola matutina*	80 650	16 218	5	5.83	0.000 <0.05
Asarum europaeum *Chelidonium majus*	96 222	31 18	5	7.81	0.000 <0.05
Viola hirta *Viola mirabilis*	136 249	33 75	9	4.16	0.001 <0.05
Viola hirta *Viola matutina*	98 502	38 108	5	8.84	0.000 <0.05
Viola hirta *Chelidonium majus*	104 332	27 165	5	3.05	0.016 <0.05
Viola mirabilis *Viola matutina*	186 908	36 245	5	6.53	0.000 <0.05
Viola mirabilis *Chelidonium majus*	256 262	34 25	5	0.32	ns
Viola matutina *Chelidonium majus*	468 366	115 81	5	1.62	ns

Thus, attractiveness of myrmecochorous diaspores for ants is determined by the following diaspore variables. At the stage of *diaspore discovery and recognition*, the principal cue is general dimension of the diaspore. At the stage of *diaspore identification*, the presence of the elaiosome and its chemical substances play the most important role in ant attraction. At the stage of *diaspore removal and transport* from the diaspore depot, elaiosome size and/or seed body curvature may be important factors for dispersal and contribute to the dispersal distance. The last hypothesis may explain results of the experiments with presentation of *Viola* sp. seeds with completely ablated elaiosomes (Imhanitskaja, 1981; Gorb and Gorb, 1995b), in which ants tried to take them due to the odour of ablated elaiosomes. However, ants were unable to transport seeds, since maximal mandible span of *F. polyctena* workers is smaller than the seed diameter (they usually hold the seed just by the elaiosome).

7. ARE THERE DIFFERENCES IN THE BODY SIZE OF ANT WORKERS WHICH REMOVE SEEDS OF DIFFERENT PLANT SPECIES?

Field observations show that *F. polyctena* workers of different size usually take diaspores of particular dimensions. Since diaspore size varies considerably among species studied, workers of different size classes would presumably prefer diaspores of different plant species. To test this hypothesis, the head width of ant workers that took diaspores of different plant species were measured. Frequency histograms of head dimensions of ants that removed seeds and of those randomly collected in the forest litter, revealed some differences between both ant groups. Frequency histograms of ants that took large seeds (*A. europaeum* and *V. hirta*), have two peaks (Fig. 2.4, A, B) and were similar to those of the sample of randomly collected ant individuals (Fig. 2.4, F). Average sizes of these three samples were identical (Fig. 2.4, G). In the case of medium-sized (*V. mirabilis*) and small-sized seeds (*V. matutina* and *Ch. majus*), the second peak is absent (Fig. 2.4, C – E), because large ants rarely removed seeds of these species. In the case of small-sized seeds (*V. matutina* and, in particular, *Ch. majus*), histograms are considerably shifted towards smaller sizes (Fig. 2.4, D, E). Workers of *F. polyctena* ants that removed *Ch. majus* seeds, were much smaller compared to the workers that removed seeds of other plant species (especially *A. europaeum* and *V. hirta*) or randomly sampled ant workers.

Large seeds (*A. europaeum, V. hirta*) were removed by *F. polyctena* ants much faster, because both elaiosome and seed body diameters correspond to the optimal mandible span of ant workers. On the contrary, elaiosomes and seed bodies of *V. matutina* and *Ch. majus* are too small for ants to manipulate with mandibles, especially large ant individuals. Therefore *F. polyctena* workers refused these seeds more frequently than seeds of *A. europaeum* and *V. hirta*. It has been previously reported for some desert ants that high foragers' polymorphism may substantially influence resource utilisation by an ant colony (Dlussky, 1981). Thus, it may be concluded that in the forest microsites with *F. polyctena* colonies with small individuals prevailing, small-seeded myrmecochores will have some advantages in diaspore dispersal. Sites with large individuals prevailing are advantageous for large-seeded myrmecochores.

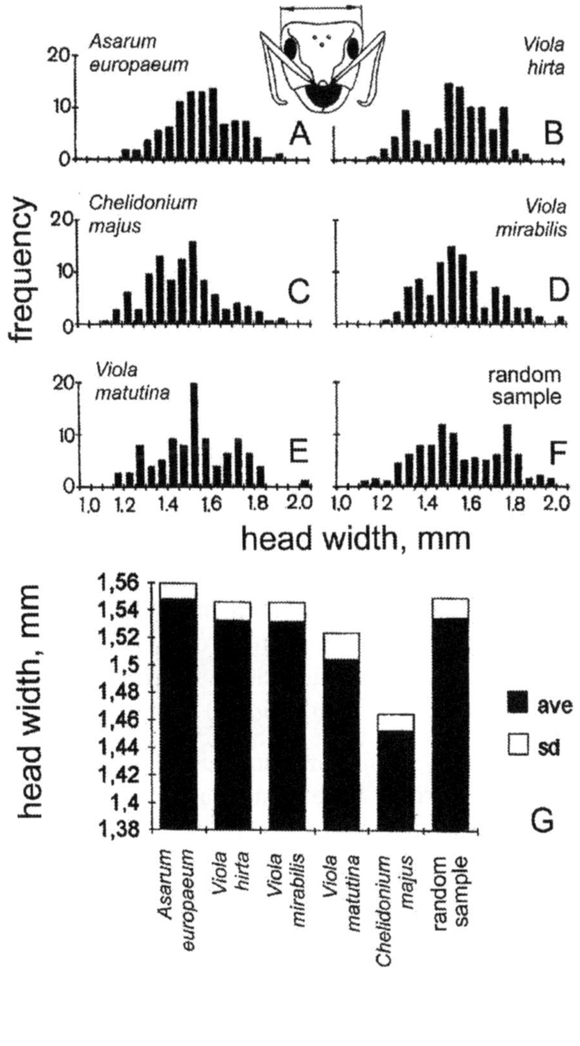

Fig. 2.4. Histograms of head width of *Formica polyctena* workers removing seeds of five species (A – E) and of randomly sampled foraging individuals (F). G. Average values (ave) and standard deviations (sd) calculated for samples presented in the histograms A – F.

Seed aggregation is a well-known factor influencing removal rates by ants (Levina, 1957; Petit et al., 1987 – 1988; Hughes and Westoby, 1990).

The exponential character of the removal dynamics curves may indicate that the removal rate may also depend on the number of seeds in the seed depot. Mobilisation reported for many ant species, including *F. polyctena* (Horstmann, 1985), *Myrmica rubra* (Cammaerts, 1978), *Odontomachus bauri* (Oliveira and Hölldobler, 1989) and others (Oliveira and Hölldobler, 1989), may greatly increase an aggregation effect during diaspore removal. A detailed role of this and other factors in seed removal is considered in detail in later chapters.

Data presented in the present chapter may lead to the conclusion that species-specific interactions between ants and myrmecochorous plants are absent (Renner, 1987; Handel and Beattie, 1990). In spite of such an absence, a complex of ant species may influence a herbaceous plant species community in the deciduous forest through different contributions of particular ant species to diaspore dispersal.

8. SUMMARY

- The first stage of myrmecochory is the removal of diaspores by ants.
- The chapter analyses interactions between the complex of plant species and the complex of ant species during removal of myrmecochorous diaspores by ants. The chapter is based mainly on results of our field experiments examining the attractiveness of seeds of obligate (*Asarum europaeum, Viola hirta*) and facultative myrmecochorous plants (*V. mirabilis, V. matutina, Chelidonium majus*) to ants of a deciduous forest.
- Seeds of different plant species are removed at different rates. Removal rates of seeds are described by exponential functions. The exponential character of removal rates was especially pronounced for obligate myrmecochores.
- Seed attractiveness increases with increasing seed size in *V. matutina, Ch. majus, V. mirabilis, V. hirta* and *A. europaeum*, and is weakly correlated with elaiosome size.
- The removal rates obtained for *Formica polyctena* ants near their nests were about ten times lower than those obtained far from the nests, where other ant species also took part in seed removal.
- Ants taking large seeds were significantly larger than ants taking small seeds. It is concluded, that the ant species complex of a deciduous forest may influence a plant species complex through seed dispersal.

Chapter 3

Factors influencing diaspore removal

In the previous chapter, it was demonstrated that diaspores of five myrmecochorous species have different attractiveness to ants of particular species and, therefore, different removal rates. Factors, such as diaspore size, or the presence and size of elaiosome or diaspore body, may effect diaspore attractiveness. In this chapter, we analyse further characters of the myrmecochorous syndrome. Data from literature and original experimental results are combined here to show the role of other factors, such as diaspore contents and diaspore aggregation in the depot for ant attraction and diaspore removal.

1. EFFECT OF THE DIASPORE SIZE IN *CORYDALIS BULBOSA* AND *CORYDALIS CAVA*

Experimental data on diaspore removal rates by ants *Formica polyctena* of five plant species clearly showed that large seeds (*Asarum europaeum, Viola hirta*) were much more attractive to ants and were removed faster by ant workers than small ones (*Chelidonium majus, V. mirabilis, V. matutina*) (see chapter 2). However, these diaspores differed not only in dimensions of elaiosomes and/or diaspore bodies, but also in diaspore shape, position of elaiosome connection to the diaspore body, and chemical content. In order to determine the effect of only diaspore size on the removal rate, an experiment with closely-related plant species is required. Such an experiment was carried out with two ephemeral early-spring species *Corydalis bulbosa* and *C. cava* and *F. polyctena* ants (for a detailed method description see chapter 11, section 4.2.3). Their seeds are similar in shape, colour, and time of fruiting, but the dimensions of the seeds and elaiosomes differ.

1.1 Morphology and anatomy of diaspores

Seeds of both species studied are shiny black, ovate, and slightly flattened at the sides (Fig. 3.1). Their diameters are significantly different ($x_{C.cava}$=2.75, sd=0.21, n=96; $x_{C.bulbosa}$=1.72, sd=0.14, n=90) (Fig. 3.1).

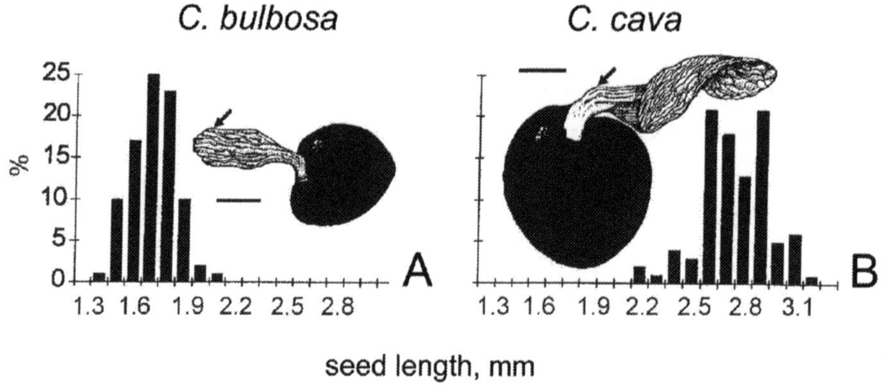

Fig. 3.1. Seeds and frequency histograms of the seed length in *Corydalis bulbosa* (A) and *C. cava* (B). Arrows indicate elaiosomes. Scale bar: 1 mm.

The elaiosome is connected to the basal part of the seed body. It is worm-like in both species. In *C. cava,* it is often twisted into a spiral (Fig. 3.1, B). Anatomical studies have revealed a funicular origin of the elaiosomes. This fact led us to consider that these appendages are true arils (for terminology see Komar (1965) and Tahtadzhjan (1985)). In longitudinal anatomical sections, it is clear that the elaiosome is composed of various cell types: the basal part consists of small round-shaped cells, the apical part – of large elongated cells. A cell group with thickened perforated cell walls is situated in the central part of the elaiosome. The last cell group presumably serves a transporting function.

1.2 Removal rates of seeds

«Cafeteria» experiments performed with seeds of *C. bulbosa* and *C. cava* in the vicinity of *F. polyctena* ants revealed that large seeds of *C. cava* were much more attractive to ants and were removed more frequently than seeds of *C. bulbosa* (Fig. 3.2). Time of half removal shows a significant difference in the plant species studied ($THR_{C.cava}$=14.31 min, sd=7.04, n=13; $THR_{C.bulbosa}$=21.43 min, sd=9.87, n=14). Thus, these experimental data are similar to our results, obtained for five other species of myrmecochores (see chapter 2), and support the hypothesis about the crucial role of seed dimensions for ant workers' attraction and, finally, for seed removal rates.

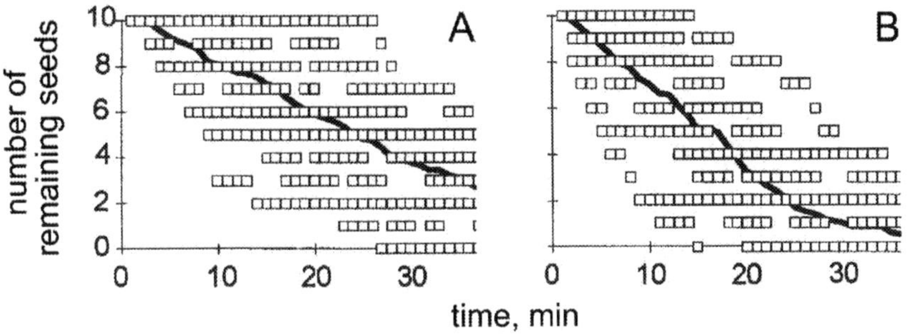

Fig. 3.2. Removal rates of seeds of *Corydalis bulbosa* (A) and *C. cava* (B) by *Formica polyctena* ants. Curves indicate average values for the complete experimental set.

It is well known that ant foraging behaviour depends on the colony polyetism (Traniello, 1989). Preferable removal of seeds of different plant species by *F. polyctena* workers of various size classes was explained by the correlation of seed diameter to the optimal mandible span of ant workers (Gorb and Gorb, 1995b). Similar results were also expected for the pair of *Corydalis* species with different seed sizes. However, difference in the head width between ant workers removing seeds of these two species was insignificant.

2. EFFECT OF ELAIOSOME, DIASPORE BODY, AND THEIR CONTENTS

The question about the significance of the elaiosomes for ant attraction was briefly addressed in chapter 2. Seeds of *Viola* sp. with elaiosomes completely ablated were not removed from the seed depot by ants. However, this may happen because of the behaviour of *F. polyctena* workers, which transport seeds by holding them in their mandibles by the elaiosome. Therefore, these experiments only partly explain the elaiosome's role in the ant attraction. To clarify the role not only of the elaiosome, but also of its content, «cafeteria» experiments were carried out with diaspores of two different types of early-spring ephemeral myrmecochores, *Corydalis cava* and *Pulmonaria obscura.* These tests were carried out with *F. polyctena* ants. The plants possess diaspores of approximately similar dimensions. The diaspores belong to different types according to their anatomy and origin. The field experiments involved intact diaspores, those with ablated elaiosomes, and «artificially» made diaspores comprising paper balls infiltrated with juices of diaspore bodies and elaiosomes (for methods see chapter 11, section 4.2.3), the role of both diaspore body and elaiosome and their content were therefore examined for ant attraction.

2.1 Seed anatomy

2.1.1 *Corydalis cava* seeds

In *C. cava*, the seed with elaiosome (Fig. 3.3, A) is a dispersal unit as in the majority of myrmecochores. Details of seed anatomy are presented in Fig. 3.3 (for methods see chapter 11, section 4.1). The basal part of the elaiosome is composed of small, almost round cells with variable wall structure. Lateral parts contain cells with dark thickened cell walls. The central part contains a group of thin-walled light cells lengthening into the seed body part adjoining the elaiosome. The middle part of the elaiosome consists of lengthened cells with thick perforated green-brownish walls. This area gradually changes into elongated cells with thin, light-coloured walls in the lateral regions of the central part of the elaiosome. The apical region of the elaiosome is composed of very large cells of irregular shape with thin walls, which become light blue after toluidine blue staining (for methods see chapter 11, section 4.1). In our preparations, these cells appeared empty.

Some authors have defined a seed appendage in *Corydalis* as a strophiole (structure in the region of the hilum or funicle) or a glandular outgrowth along the raphe (Sernander, 1906) without considering its relation to the aril. Other authors thought that the seed appendage in *Corydalis* is a variant of the aril (Tahtadzhjan, 1964). Later anatomical studies considered the strophiole as a derivation of the external integument along the line raphe-antiraphe, and that the true aril is a structure emerging on the funicle and not joining with the seed body (Komar, 1965; Tahtadzhjan, 1985). Our data have supported the funicular origin of the seed appendage in *Corydalis*. Hence, it may be determined as a rue aril and not a strophiole. Absence of the cell content in representatives of the genus *Corydalis* was previously described by Pfeiffer (1891), who considered seed appendages in *Asarum, Viola*, and *Corydalis* as storing tissues. Other authors have reported rich fatty inclusions in transparent cells of *Corydalis* arils (Tahtadzhjan, 1985). This is the reason why they are often just called elaiosomes (Komar, 1965; Sernander, 1906), adaptations to ant-dispersal of seeds that were found in a variety of representatives of the plant family Fumariaceae (Berg, 1966; 1969; Beattie, 1983).

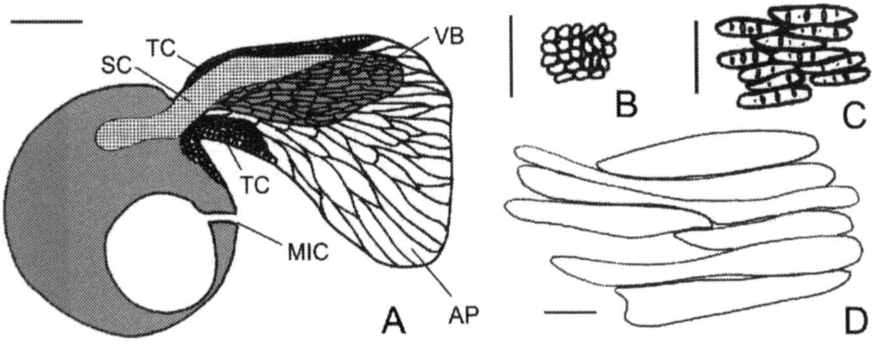

Fig. 3.3. Anatomy of the seed of *Corydalis cava*. A. Diagram of a dissected seed. B. Small cells with thin walls. C. Cells of the supporting tissue. D. Cells of the apical part of the elaiosome. AP, apical part of the elaiosome; MIC, micropyle; VB, vascular bundle; SC, small cells with thin walls; TC, cells with thick walls. Scale bars: A – 0.45 mm; B, C – 0.20 mm; D – 0.05 mm.

Generative diaspores, adapted to ant-dispersal, usually bear a rather solid integument to avoid damage by ant workers. However, *Corydalis* seeds have relatively thin spermoderm which may be easily damaged by ant mandibles during manipulation of seeds. Anatomical studies have revealed a group of small thick-walled cells at the base of the elaiosome. Presumably, these cells (Fig. 3.3, A, B) provide easy breakage of the elaiosome and thus a decrease in the time of manipulation with the seed after its transport from the seed depot. This adaptation may effect a decrease of probability in seed damage in *C. cava*.

2.1.2 *Pulmonaria obscura* erems

The fruit of *P. obscura* is a coenobium in an effused flower bell, which disintegrates into separate erems. In this case, the dispersal unit is the erem (Fig. 3.4, A). Erems of *P. obscura* are rather large, ovate, with a pointed apical part, smooth, shiny and dark-brownish. In its basal part, the erem has a collar-shaped thickening that surrounds a small round whitish elaiosome. Erem length is much less variable (x_{ave}=4.44 mm, sd=0.57, n=100) compared to the elaiosome length (x_{ave}=0.74 mm, sd=0.23, n=100).

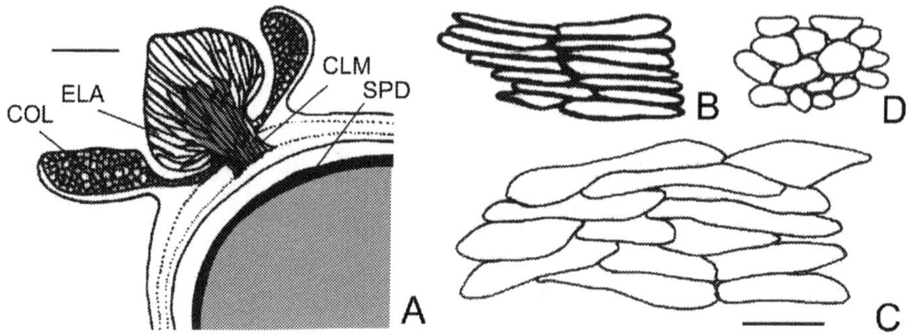

Fig. 3.4. Anatomy of the erem of *Pulmonaria obscura*. A. Diagram of a dissected erem. B. Column cells. C. Cells of fundamental tissue of elaiosome. D. Parenchymatous cells of the collar. CLM, column; COL, collar; ELA, elaiosome; SPD, spermoderm. Scale bar: A – 0.45 mm; B – D – 0.08 mm.

Details of the anatomy of both erems and elaiosomes of *P. obscura* are shown in Fig. 3.4 (for methods see chapter 11, section 4.1). The elaiosome is a fruit (erem) appendage emerging in the place of its attachment to the pedicle. The elaiosome develops from middle and internal layers of the pericarp., The so-called column is located at the place of appendage attachment. This is a small group of tightly-packed, elongated, narrow cells, which are brightly-purple coloured after staining with toluidine blue. The cells presumably serve a transporting function. In its upper part, beyond the pericarp, the column is gradually transformed into narrowly-elongated cells, coloured bluish-purple (less intensively than column cells) by the toluidine blue stain. The main part of the elaiosome (apical and lateral regions) consists of large, thin-walled, elongated, widened cells, coloured light blue by toluidine blue. In our preparations the cells appeared empty, without content.

The external cell layer of the fruit coat is covered by a thin layer of non-cellular material. It is stained dark-purple by toluidine blue. The collar is composed of deflected regions of the exocarp, which are covered by several layers of round or oval cells on the internal side (facing the elaiosome). The cells have thin walls, coloured purple by toluidine blue stain. At the edge of the collar, the number of layers increases. In the medial part of the erem, layers of parenchymatous cells of the collar are connected to the elaiosome.

It has been previously mentioned that, in representatives of the plant families Boraginaceae and Lamiaceae, fruit structures are adapted for ant-dispersal (Sernander, 1906; Ulbrich, 1928; Bresinsky, 1963). Fat-containing elaiosomes of *P. obscura* fruits attract ants in a similar way as seed elaiosomes of other plants. The basal location of the elaiosome is convenient for handling by ant workers. Sernander (1906) considered the elaiosome of *P. officinalis* to be a false strophiole (pseudostrophiole) according to data of earlier authors. Our anatomical data demonstrate that the elaiosome of *P. obscura* originates from the endo- and exocarp in the place of the erem joint to the receptacle, and, therefore, cannot be attributed to the strophiole. Thus, diaspores of *C. cava* and *P. obscura* are analogous structures, convergently appearing in the evolution of these plant lineages.

2.2 Experiments with diaspores and juices from elaiosomes and diaspore bodies

Intact diaspores of myrmecochores studied have similar removal rates by *F. polyctena* ants (Fig. 3.5, B) and the time of half removal does not significantly differ (Fig. 3.5, A).

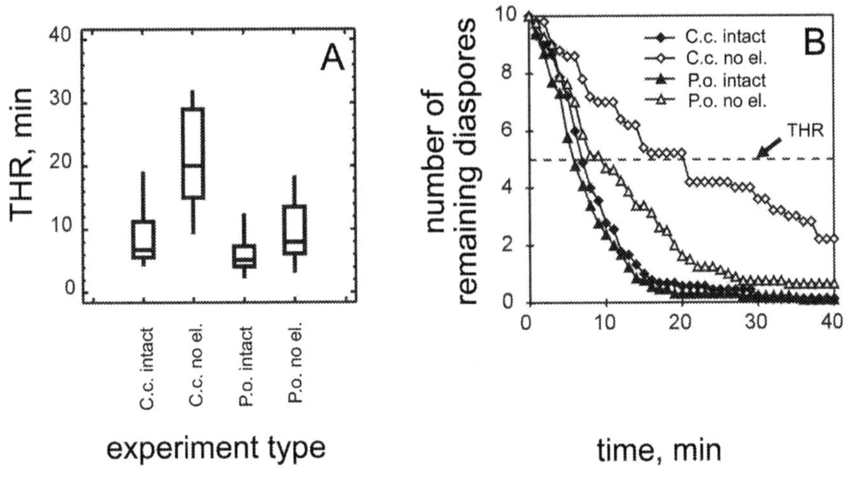

Fig. 3.5. Time of half removal (A) and removal rates (B) of diaspores of *Corydalis cava* and *Pulmonaria obscura* in experiments with elaiosomes ablated. C.c. intact, intact seeds of *C. cava*; C.c. no el., seeds of *C. cava* without elaiosomes; P.o. intact, intact diaspores of *P. obscura*; P.o. no el., diaspores of *P. obscura* without elaiosomes; THR, time of half removal.

2.2.1 *Corydalis cava* seeds

Experiments with removal of *C. cava* seeds with elaiosomes ablated, and with paper balls, infiltrated with juices of seed bodies and elaiosomes, showed the crucial role of elaiosome content in the ant attraction. Ant workers removed intact seeds and balls with elaiosome juice more frequently than control balls infiltrated with distilled water or seed body juice (Fig. 3.6, B). Time of half removal of intact seeds and balls with elaiosome juice did not significantly differ (Fig. 3.6, A). In another series of experiments, seeds with elaiosomes ablated were removed significantly slower than intact seeds (Fig. 3.5, B). The time of half removal of intact seeds (x_{ave}=8.9 min, sd=4.9, n=12) was significantly lower than those with elaiosomes ablated (x_{ave}=21.1 min, sd=9.5, n=5) (Fig. 3.6, A).

Seed body content of *C. cava* seeds was not attractive to ant individuals. Time of half removal of balls with seed body juice and with distilled water did not significantly differ (Fig. 3.6, A). Since the anatomical study of diaspores did not reveal a lipid-containing layer on the seed body surface, weak attractiveness of *C. cava* seed bodies may be explained by an absence of specific adaptations of seed bodies in this species to ant attraction.

2.2.2 *Pulmonaria obscura* erems

Intact diaspores of *P. obscura* were removed significantly faster than paper balls infiltrated with distilled water, elaiosome juice, and erem juice (Fig. 3.6, D). Time of half removal of balls with juices of erems and elaiosomes did not significantly differ (Fig. 3.6, C). Time of half removal of balls with erem or elaiosome juice was significantly lower than that of balls infiltrated with distilled water (Fig. 3.6, C).

Removal rates of diaspores with elaiosomes ablated was slightly higher than that of intact diaspores (Fig. 3.5, B). Time of half removal (9.3 min, sd=5.1, n=8 and 6.1 min, sd=3.5, n=13, respectively) did not significantly differ (Fig. 3.5, A). Such similarity in attractiveness of intact diaspores and those with elaiosomes ablated may be explained by the presence of additional adaptations of diaspores of *P. obscura* to the myrmecochory. Presumably, the fat-containing erem coat and fat-containing tissues of the internal surface of the collar attract ants even after elaiosome ablation. In contrast to parenchymatous cells of elaiosome, which can be easily bitten off by ants, the pericarp is rather dense and thus protects the erem from damage by ant individuals. The presence of a collar-shaped thickening makes erem seizure and transport by ants much easier, if the elaiosome is absent.

2.2.3 Diaspore structures that attract ants

It is known that the key biochemical stimuli responsible for ant attraction and for the release of ant carrying behaviour are lipid substances (see chapter 1, section 3.4). Results of experiments with paper balls, filled with juices, provide the evidence that chemical releasers may be located in different parts of the diaspore, not necessarily in the elaiosome. Ants responded similarly to balls with the elaiosome juice and to intact seeds of *C. cava*, and weakly to balls with seed body juice. Thus, elaiosome content alone can be fully responsible for seed attraction to ants for this plant species.

Surprisingly, juices of both erems and elaiosomes of *P. obscura* were equally attractive to ant workers. Presumably, contribution of collar cells' and external coat contents to the entire content of the erem juice promote attractiveness of the erem. However, each juice was less attractive than intact diaspores. It seems that the simultaneous presence of elaiosome and erem components makes intact erems much more attractive than their juices presented separately.

Furthermore, the presence of visual releasers, contributing to the diaspore removal from the litter, cannot be excluded. Delpino believed that erem contour, colour, and pubescence resemble posterior parts of the body of the ant *Camponotus pubescens* (Sernander, 1906). Diaspore appearance in *P. obscura* is very similar to the metathorax and abdomen of some curculionid beetles. These often occur as remains in samples of the mound material of *F. polyctena* ants' nests and soil samples from territory borders of this ant species (Gorb and Gorb, 1998b; Gorb et al., 2000).

It was previously demonstrated that diaspore removal rates by *F. polyctena* ants depend more on diaspore dimensions and less on elaiosome size (see chapter 2). Wide variability of elaiosome size compared to the erem size in *P. obscura* supports the hypothesis that the size of the entire diaspore is a more important factor for the ant attraction and diaspore removal than the size of the elaiosome (Gorb and Gorb, 1995b). Since *P. obscura* pericarp is attractive for ants, it is capable of providing a diaspore dispersal in the case of a very small elaiosome or even without it. If so, it is clear why elaiosome size in this plant species is weakly controlled by natural selection. In contrast to the elaiosome, erem size is, presumably, under control of a stabilised selection. A low variability of the erem size may be explained by the high importance of this character in providing an optimal ratio between ease of diaspore finding (large diaspore) and ease of its transporting (small diaspore) by ants. Shape and size of *P. obscura* diaspores aid in the ant attraction and in contribution not only to the primary, but also to the secondary diaspore removal after being dropped during transport by ants

(Gorb and Gorb, 1999a). Nevertheless, other hidden factors may also influence control of the diaspore size by natural selection.

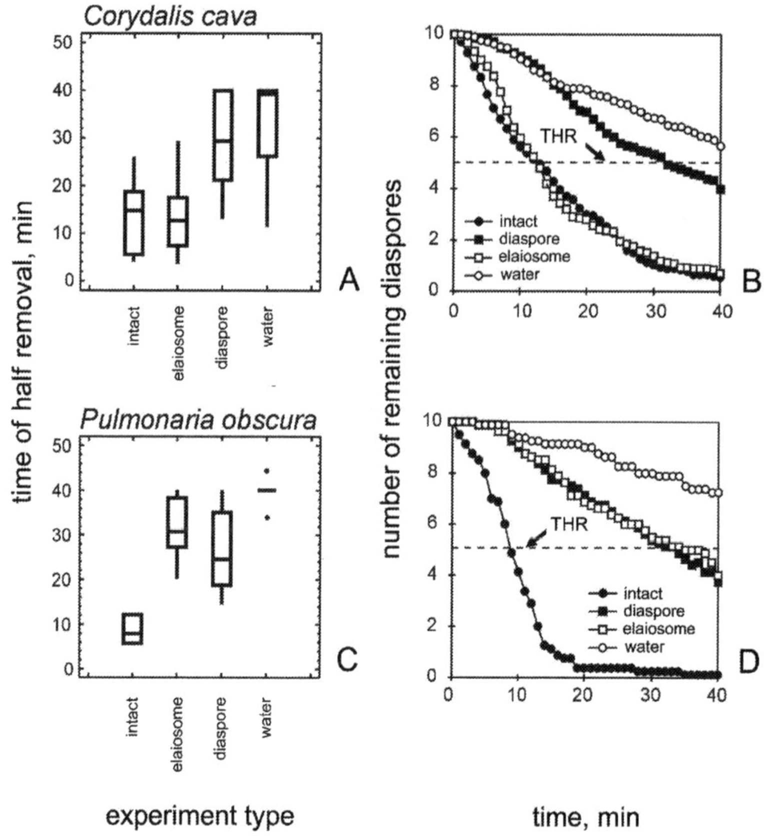

Fig. 3.6. Time of half removal (A, C) and removal rates (B, D) of diaspores of *Corydalis cava* (A, B) and *Pulmonaria obscura* (C, D) by *F. polyctena* ants in experiments with paper balls filled with elaiosome and diaspore body juices. diaspore, extract of diaspore bodies; elaiosome, juice of elaiosomes; intact, intact diaspores; THR, time of half removal; water, distilled water (control).

3. EFFECT OF DIASPORE AGGREGATION

The fact that larger diaspores are removed much faster by ant workers was explained by a higher chance of detection by ant foragers. Larger seeds are more visible and have a much stronger odour. If this is so, the degree of diaspore aggregation may enhance their being found and removed by ant foragers. It is well known that foragers respond differently to one large food object than to a group of small objects (Bisseau and Pasteels, 1994). It was also previously reported from field observations that intensity of seed removal directly depends on their number in the group (Levina, 1957). The exponential character of curves of removal dynamics in myrmecochorous plants may be also explained by different numbers of diaspores in the depot at the beginning and at the end of the experiment. Effect of diaspore aggregation on removal rates has been previously reported for ant species with mobilisation behaviour for food objects (Hughes and Westoby, 1990). However, removal of *Carex pilulifera* diaspores by *Myrmica ruginodis* ants was not influenced by the degree of diaspore aggregation (Kjellsson, 1985).

3.1 «Cafeteria» experiments with different degrees of seed aggregation in the depot

The size of the food object can be simulated in a field experiment by various degrees of aggregation of the elaiosome-bearing diaspores. In reality, diaspore aggregation probably plays a more important role for the diaspore removal of obligate myrmecochores, since their diaspores are usually accessible in groups to foraging ants. This hypothesis was examined in experiments on seed removal from seed depots, in which seeds were located at various distances from one another.

Diaspores of two plant species (obligate myrmecochore *Viola odorata* with large seeds and diplochore *Chelidonium majus* with small seeds) and one ant species (*Formica polyctena*) were used in experiments. Registration of the seed number versus time were made simultaneously on six seed depots, in which seeds were located at the same distance from one another (0, 1, 2, 4, 8, 16 mm) (for methods see chapter 11, section 4.2.3).

Seeds of *V. odorata* are considerably larger than seeds of *Ch. majus* (Table 1.1), and have higher removal rates by *F. polyctena* ants (Fig. 3.7). Aggregated and solitary seeds of *V. odorata* were removed from depots by ants at the same rate. Curves of the removal dynamics of *V. odorata* seeds with various degrees of aggregation did not significantly differ (Fig. 3.7, A). Values of the second exponential coefficient, describing removal rate, were very similar for all six degrees of aggregation (Table 3.1).

The difference between results, obtained on different seed depots, was more distinct, if the time of removal of the fifth seed was compared, and not the first seed (Fig. 3.7, C). The time of half removal was also not significantly different between various degrees of aggregation.

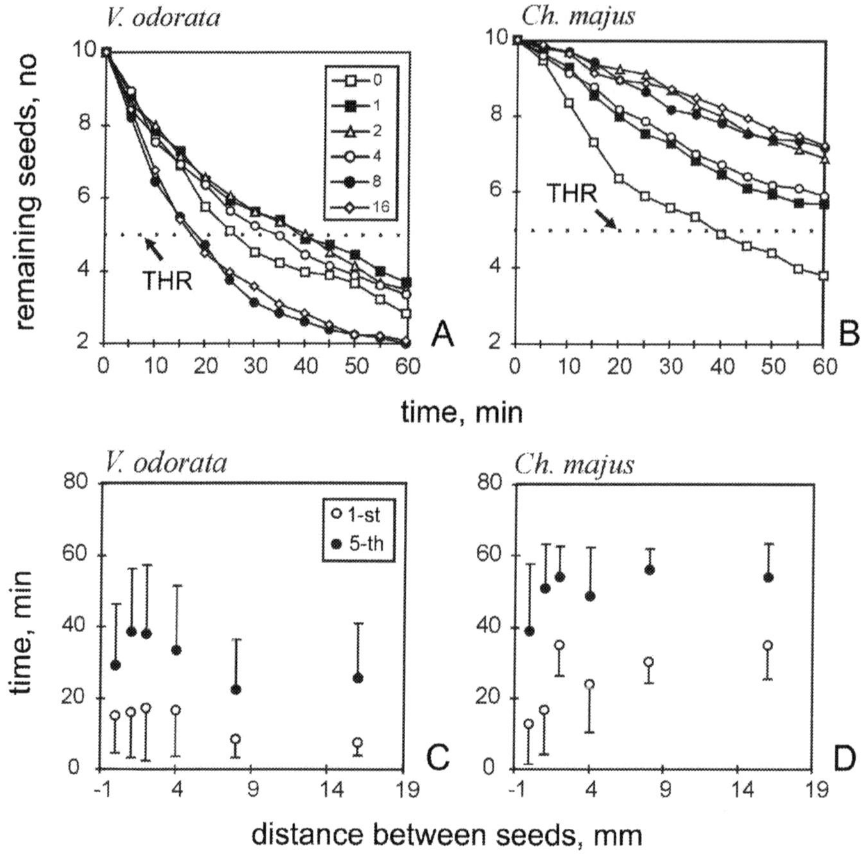

Fig. 3.7. Effect of seed aggregation on the removal rates of *Viola odorata* and *Chelidonium majus* seeds by *F. polyctena* ants. A, B. Removal rates of seeds deposited in six patterns in depressions located at different distances (0, 1, 2, 4, 8, and 16 mm). C, D. Dependence of the time of removal of the first and fifth seeds versus distance between seeds in the depot. THR, time of half removal.

Aggregated seeds of *Ch. majus* were removed faster than non-aggregated ones. The types of curves, describing removal dynamics, was considerably different (Fig. 3.7, B). Seed removal was faster on experimental tables, where the seeds were aggregated more densely. In other words: the larger the distance between seeds in the depot, the lower the seed removal rate. The slowest removal was registered for seeds separated by 2, 8, and 16 mm. The second exponential coefficient demonstrated such a difference quite well (Table 3.1).

In contrast to the results with *V. odorata*, the difference between depots with various degrees of aggregation was more distinct, if removal time of the first, not the fifth, seed was considered (Fig. 3.7, D). This value indicates time of finding of a seed group or just a single seed by ants. The result may be explained by the small size of seeds of this plant species. Single small seeds of *Ch. majus* are weakly visible for foraging ants, and time of detection of the seed group becomes an important factor effecting removal rates.

Table 3.1. Average values of the second exponential coefficient (*b*) ($y=a\cdot exp(-b\cdot x)$), describing removal rates of seeds.

Distance between seeds, mm	*Viola odorata*	*Chelidonium majus*
0	$4.60\cdot10^{-2}$	$2.57\cdot10^{-2}$
1	$3.24\cdot10^{-2}$	$1.45\cdot10^{-2}$
2	$3.85\cdot10^{-2}$	$0.78\cdot10^{-2}$
4	$3.31\cdot10^{-2}$	$1.47\cdot10^{-2}$
8	$5.11\cdot10^{-2}$	$0.71\cdot10^{-2}$
16	$5.18\cdot10^{-2}$	$0.85\cdot10^{-2}$

3.2 Diaspore aggregation in obligate and facultative myrmecochores

A single seed of *V. odorata* is large and therefore perceptible for foragers of *F. polyctena*, which are usually attracted by large food objects. Non-aggregated seeds of this species were not significantly more attractive, because they covered a relatively large area, and, consequently, a larger number of ants crossed this area. Thus, large single seeds of obligate myrmecochores, in contrast to small seeds of facultative myrmecochores, have a better chance to be secondarily removed by ant individuals after seeds get dropped in the forest litter by other ants transporting them (Gorb and Gorb, 1999a).

In contrast to *V. odorata* seeds, a single seed of *Ch. majus* is small and hardly noticeable by ants. They usually passed by such a seed. However, aggregated seeds of this myrmecochore look like a large food object, and, therefore, may be easily detected by foragers of ant species, as large as *F. polyctena*. However, in nature, *Ch. majus* seeds are not aggregated; they are usually scattered in the litter after the first (autochorous) stage of dissemination. It can be suggested that seeds of small-seeded diplochores are not adapted for recognition and removal by large ant species, but rather by small-sized species, such as *Myrmica rubra, Lasius flavus, L. fuliginosus, L. niger*, etc.

Of course, seed aggregation may also cause some disadvantages for the plant. Aggregated seeds may be localised much faster by potential seed predators, such as rodents (O'Dowd and Hay, 1980; Heithaus, 1981; Turnbull and Culver, 1983; Gibson, 1993b) and ground beetles (Ohara and Higashi, 1987; Higashi and Ito, 1991; Espadaler and Gomez, 1996; Ohkawara et al., 1996; 1997). Seeds of diplochorous species escape from predators due to their small size and an additional stage of seed dispersal (usually various types of autochory) (Ohkawara and Higashi, 1994). Seeds of obligate myrmecochores, due to their large size and dense aggregation, are very rapidly removed by ants and also have a high chance of escaping potential predators. Thus, obligate myrmecochory and diplochory are two alternative ant-dispersal strategies, providing dispersal for distance and escape from seed predators. They differ by a combination of (1) dispersal stages, (2) seed and elaiosome sizes, (3) degrees of seed aggregation, at which seeds are accessible to ant workers.

4. SUMMARY

- In this chapter, the effect of some characters of the myrmecochorous syndrome on the removal rates of diaspores by ants is analysed. The study of the morphology and dispersal of seeds of two closely-related species of *Corydalis* showed that large seeds of *C. cava* were more attractive to *Formica polyctena* ants and were removed more rapidly than small seeds of *C. bulbosa*.

- Field experiments included diaspores and juices of diaspore parts of *C. cava* and *Pulmonaria obscura* to test the effect of morphological and anatomical adaptations of diaspores and elaiosomes on diaspore attractiveness to *F. polyctena* ants. It was shown that chemical cues attracting ants are located in different parts of the diaspore: in the elaiosome (*C. cava, P. obscura*), in the coverage of the external layer of the fruit, and in the fruit collar (*P. obscura*).

- An investigation examined the effect of seed aggregation on seed removal rates by *F. polyctena* ants in a small-seeded diplochore *Chelidonium majus* and a large-seeded obligate myrmecochore *Viola odorata*. In *Ch. majus* the removal rates depended significantly on seed aggregation: seed removal was quicker from depots where seeds were closely aggregated. There were no significant differences in removal

Chapter 4

Effect of the ant species complex on diaspore removal

Diaspores of myrmecochorous plants are usually removed not by one but rather several different ant species. A. Beattie (1983) provided a list of 80 ant taxa, which were registered in the literature as seed dispersal agents. Ant individuals of the same species usually remove diaspores of different plant species at various rates, depending on the diaspore size and other characteristics of the myrmecochorous syndrome, as well as on the size of the ant individuals (Culver and Beattie, 1980; Nesom, 1981; Clay, 1983; Wein and Pickett, 1989; Higashi et al., 1989; Gorb and Gorb, 1995a; b; Gorb, 1998). Ant species can have different life histories and behaviour, which may influence diaspore removal rates, seed flow on ant territories, and seed concentration within ant nests to a different extent. Furthermore, foraging behaviour of the same ant species may change depending on the presence of other species in the ecosystem. It is suggested here that not just one species but rather it is myrmecocomplexes that have an effect on diaspore dispersal and on the distribution of seedlings. Such myrmecocomplexes also influence plant species complexes in the ecosystem. In this chapter, experimental results of seed removal of the violet *Viola odorata* by four ant species in three forest microsites, differing in ant species composition, are presented.

To estimate the contribution of foragers of different ant species (*Formica polyctena* (FP), *Lasius fuliginosus* (LF), *Myrmica rubra* (MR), and *Leptothorax nylanderi* (LN)) (Fig. 4.1) to seed removal, the following comparisons were made: (1) number of visits of experimental seed depots by ants and seed removal rates in three microsites (*F. polyctena* territory: complex FP+MR+LN; *L. fuliginosus* territory: complex LF+MR+LN; *M. rubra* territory: complex MR+LN); (2) seed removal dynamics by ants of different species; (3) dependence of the duration of manipulation of ants with seeds on the microsite type; (4) number of seeds, to which ant workers of different species have had contact prior to seed removal; (5) contribution of the recruitment and learning in different ant species to removal (for a detailed method description see chapter 11, section 4.2.4).

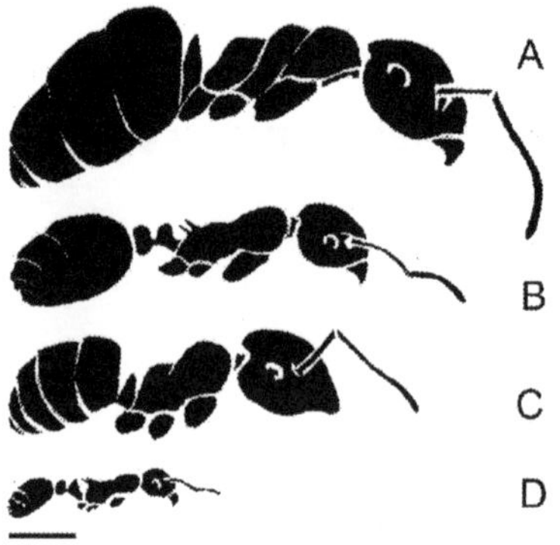

Fig. 4.1. Ant species used in experiments. A. *Formica polyctena*. B. *Myrmica rubra*. C. *Lasius fuliginosus*. D. *Leptothorax nylanderi*. Scale bar: 1 mm.

1. VISITS TO SEED DEPOTS BY ANT FORAGERS AND SEED REMOVAL

1.1 Total seed removal at different microsites

At microsites with different ant species composition, the number of visits to seed depots by foragers was significantly different (Fig. 4.2, A). In *L. fuliginosus* territory, seed depots were visited more frequently than in territories of both *F. polyctena* and *M. rubra*. Seed removal rates were also different between microsites. Visits to depots by ants were more often accompanied by seed removal in the territories of *F. polyctena*, and *L. fuliginosus* (Fig. 4.2, B). Seed removal by *L. nylanderi* ants was very rarely observed at all three microsites. *M. rubra* workers had different ratios of removal-to-refusal events at various microsites.

After four hours exposure of seed depots to ants at the microsite with the complex FP+MR+LN, about 100 seeds were totally removed and transported by ants. At the microsites with complexes of MR+LN and LF+MR+LN, only 20 and 40 seeds, respectively, were removed (Fig. 4.3). The fastest removal occurred at the microsite FP+MR+LN (tangent of the slope of the linear regression line is 0.4324), the slowest one – at microsites LF+MR+LN (0.1595), and MR+LN (0.0892). Removal rate was constant over the exposure time at the microsite MR+LN, decreased at the microsite MR+LN and increased at the microsite LF+MR+LN (Fig. 4.4).

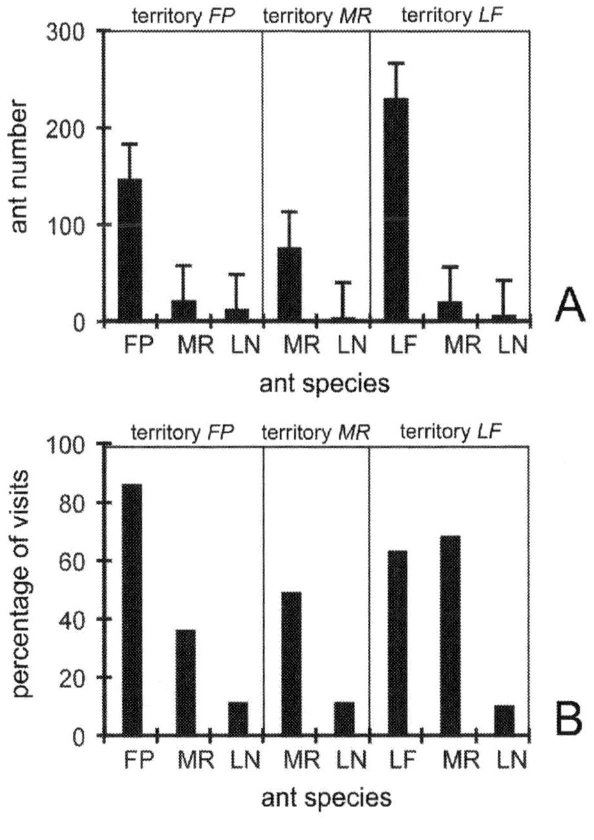

Fig. 4.2. Ant visits (A) and total seed removal (B) summarised for microsites with different ant species compositions in experiments with direct observations of the seed depot. FP, *Formica polyctena*; LF, *Lasius fuliginosus*; LN, *Leptothorax nylanderi*; MR, *Myrmica rubra*; X-axis, ant species according to microsites; Y-axis (top), number of ants that visited seed depot; Y-axis (bottom), percentage of ant individuals that removed seeds.

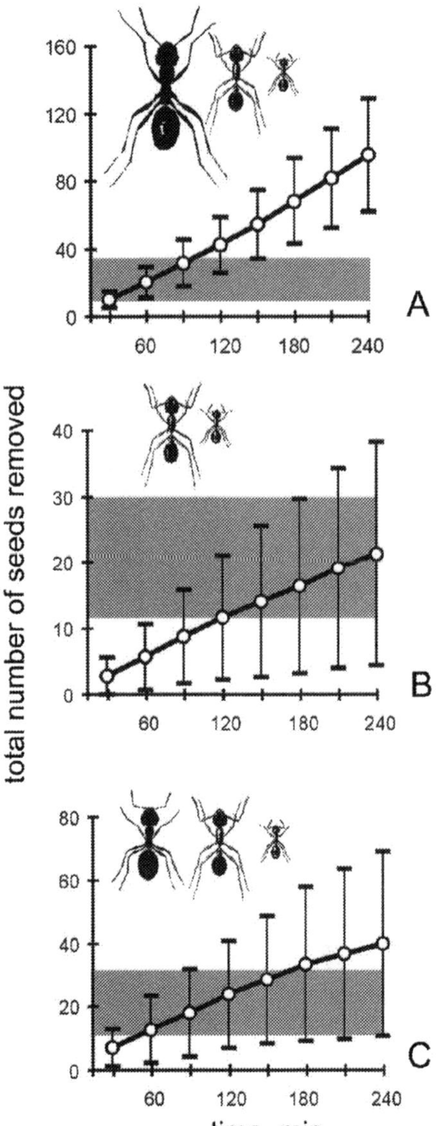

Fig. 4.3. Cumulative removal (total number of seeds removed) in the *Formica polyctena* (A), *Myrmica rubra* (B) and *Lasius fuliginosus* (C) territories. Grey bands indicate areas of 10 – 35 seeds, which usually correspond to the fruit volume of *Viola odorata*. Tangent of the linear regression line corresponds to the number of seeds removed in 1 min.

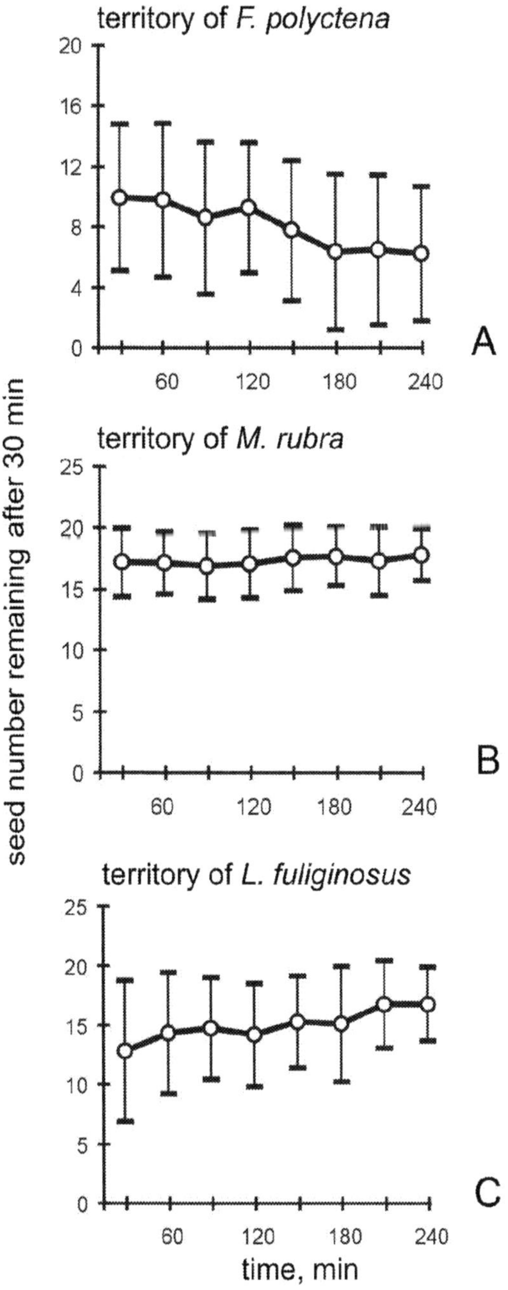

Fig. 4.4. Seed removal over 30 min time intervals at different microsites.

1.2 Contribution of different ant species to seed removal

1.2.1 *Formica polyctena* territory

Ant foragers of the dominating species (*F. polyctena*) mainly visited seed depots and removed seeds at this microsite (Fig. 4.2). Various size classes of *F. polyctena* workers, visually determined as large, middle-sized, and small, were characterised by similar removal-to-refusal ratios (large: 88% and 12%, n=84; middle-sized: 87% and 13%, n=627; small: 85% and 15%, n=212, respectively). Tangents of linear regression lines for *F. polyctena* workers only and for all three ant species of the microsite were similar for visits (0.0927 and 0.0892, respectively), as well as for seed removal (0.0779 and 0.0751, respectively).

In the cases of *M. rubra* and *L. nylanderi* workers, both visiting and removal dynamics remain more-or-less constant during the depot exposure to ants. For *M. rubra*, tangents of regression lines were -0.0092 (visits) and -0.0023 (removal). For *L. nylanderi*, they were 0.0009 and 0.0002, respectively. For *F. polyctena*, removal dynamics were clearly correlated with the visiting dynamics, because the number of refusals was rather low (15%, n=1163). For *M. rubra* and *L. nylanderi*, the number of refusals was considerably higher (*M. rubra*: 64%, n=157; *L. nylanderi*: 89%, n=90) (Fig. 4.2, B).

1.2.2 *Myrmica rubra* territory

M. rubra workers mainly contributed to both the depot visits and seed removal (Fig. 4.2). Removal rates by both ant species, inhabiting this microsite, were constant during the exposure time. Tangents of linear regression lines for *M. rubra* only and for both ant species were rather similar for visits (0.0089 and 0.0079, respectively) and for removal (0.0033 and 0.0037, respectively). *L. nylanderi* workers regularly visited seed depots, but very rarely removed seeds. Removal was observed only in 11% of visits (n=27). Tangents of the linear regression line slope were 0.0003 (visits) and -0.0002 (removal).

1.2.3 *Lasius fuliginosus* territory

At this microsite, exposure time of seed depots to ants was shorter (1.5 h instead of 4 h), because of a high visiting rate of *L. fuliginosus* workers that made observation of depots and data recording difficult. The dominant species (*L. fuliginosus*) contributed mainly to the overall number of visits

Chapter 4

Effect of the ant species complex on diaspore removal

Diaspores of myrmecochorous plants are usually removed not by one but rather several different ant species. A. Beattie (1983) provided a list of 80 ant taxa, which were registered in the literature as seed dispersal agents. Ant individuals of the same species usually remove diaspores of different plant species at various rates, depending on the diaspore size and other characteristics of the myrmecochorous syndrome, as well as on the size of the ant individuals (Culver and Beattie, 1980; Nesom, 1981; Clay, 1983; Wein and Pickett, 1989; Higashi et al., 1989; Gorb and Gorb, 1995a; b; Gorb, 1998). Ant species can have different life histories and behaviour, which may influence diaspore removal rates, seed flow on ant territories, and seed concentration within ant nests to a different extent. Furthermore, foraging behaviour of the same ant species may change depending on the presence of other species in the ecosystem. It is suggested here that not just one species but rather it is myrmecocomplexes that have an effect on diaspore dispersal and on the distribution of seedlings. Such myrmecocomplexes also influence plant species complexes in the ecosystem. In this chapter, experimental results of seed removal of the violet *Viola odorata* by four ant species in three forest microsites, differing in ant species composition, are presented.

To estimate the contribution of foragers of different ant species (*Formica polyctena* (FP), *Lasius fuliginosus* (LF), *Myrmica rubra* (MR), and *Leptothorax nylanderi* (LN)) (Fig. 4.1) to seed removal, the following comparisons were made: (1) number of visits of experimental seed depots by ants and seed removal rates in three microsites (*F. polyctena* territory: complex FP+MR+LN; *L. fuliginosus* territory: complex LF+MR+LN; *M. rubra* territory: complex MR+LN); (2) seed removal dynamics by ants of different species; (3) dependence of the duration of manipulation of ants with seeds on the microsite type; (4) number of seeds, to which ant workers of different species have had contact prior to seed removal; (5) contribution of the recruitment and learning in different ant species to removal (for a detailed method description see chapter 11, section 4.2.4).

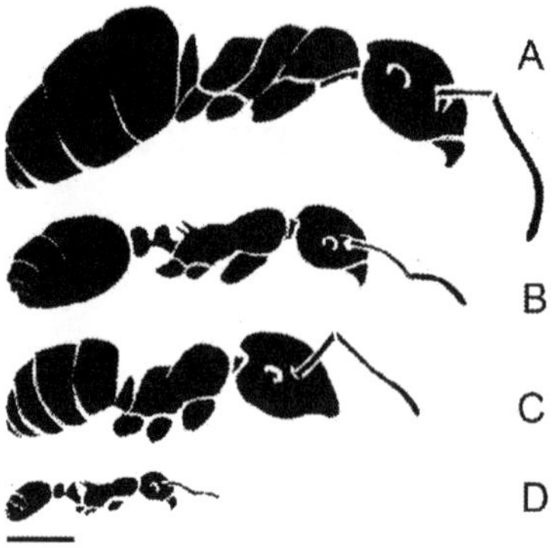

Fig. 4.1. Ant species used in experiments. A. *Formica polyctena*. B. *Myrmica rubra*. C. *Lasius fuliginosus*. D. *Leptothorax nylanderi*. Scale bar: 1 mm.

1. VISITS TO SEED DEPOTS BY ANT FORAGERS AND SEED REMOVAL

1.1 Total seed removal at different microsites

At microsites with different ant species composition, the number of visits to seed depots by foragers was significantly different (Fig. 4.2, A). In *L. fuliginosus* territory, seed depots were visited more frequently than in territories of both *F. polyctena* and *M. rubra*. Seed removal rates were also different between microsites. Visits to depots by ants were more often accompanied by seed removal in the territories of *F. polyctena*, and *L. fuliginosus* (Fig. 4.2, B). Seed removal by *L. nylanderi* ants was very rarely observed at all three microsites. *M. rubra* workers had different ratios of removal-to-refusal events at various microsites.

After four hours exposure of seed depots to ants at the microsite with the complex FP+MR+LN, about 100 seeds were totally removed and transported

In contrast to the results with *V. odorata*, the difference between depots with various degrees of aggregation was more distinct, if removal time of the first, not the fifth, seed was considered (Fig. 3.7, D). This value indicates time of finding of a seed group or just a single seed by ants. The result may be explained by the small size of seeds of this plant species. Single small seeds of *Ch. majus* are weakly visible for foraging ants, and time of detection of the seed group becomes an important factor effecting removal rates.

Table 3.1. Average values of the second exponential coefficient (b) ($y=a\cdot exp(-b\cdot x)$), describing removal rates of seeds.

Distance between seeds, mm	*Viola odorata*	*Chelidonium majus*
0	$4.60\cdot10^{-2}$	$2.57\cdot10^{-2}$
1	$3.24\cdot10^{-2}$	$1.45\cdot10^{-2}$
2	$3.85\cdot10^{-2}$	$0.78\cdot10^{-2}$
4	$3.31\cdot10^{-2}$	$1.47\cdot10^{-2}$
8	$5.11\cdot10^{-2}$	$0.71\cdot10^{-2}$
16	$5.18\cdot10^{-2}$	$0.85\cdot10^{-2}$

3.2 Diaspore aggregation in obligate and facultative myrmecochores

A single seed of *V. odorata* is large and therefore perceptible for foragers of *F. polyctena*, which are usually attracted by large food objects. Non-aggregated seeds of this species were not significantly more attractive, because they covered a relatively large area, and, consequently, a larger number of ants crossed this area. Thus, large single seeds of obligate myrmecochores, in contrast to small seeds of facultative myrmecochores, have a better chance to be secondarily removed by ant individuals after seeds get dropped in the forest litter by other ants transporting them (Gorb and Gorb, 1999a).

In contrast to *V. odorata* seeds, a single seed of *Ch. majus* is small and hardly noticeable by ants. They usually passed by such a seed. However, aggregated seeds of this myrmecochore look like a large food object, and, therefore, may be easily detected by foragers of ant species, as large as *F. polyctena*. However, in nature, *Ch. majus* seeds are not aggregated; they are usually scattered in the litter after the first (autochorous) stage of dissemination. It can be suggested that seeds of small-seeded diplochores are not adapted for recognition and removal by large ant species, but rather by small-sized species, such as *Myrmica rubra, Lasius flavus, L. fuliginosus, L. niger*, etc.

Of course, seed aggregation may also cause some disadvantages for the plant. Aggregated seeds may be localised much faster by potential seed predators, such as rodents (O'Dowd and Hay, 1980; Heithaus, 1981; Turnbull and Culver, 1983; Gibson, 1993b) and ground beetles (Ohara and Higashi, 1987; Higashi and Ito, 1991; Espadaler and Gomez, 1996; Ohkawara et al., 1996; 1997). Seeds of diplochorous species escape from predators due to their small size and an additional stage of seed dispersal (usually various types of autochory) (Ohkawara and Higashi, 1994). Seeds of obligate myrmecochores, due to their large size and dense aggregation, are very rapidly removed by ants and also have a high chance of escaping potential predators. Thus, obligate myrmecochory and diplochory are two alternative ant-dispersal strategies, providing dispersal for distance and escape from seed predators. They differ by a combination of (1) dispersal stages, (2) seed and elaiosome sizes, (3) degrees of seed aggregation, at which seeds are accessible to ant workers.

4. SUMMARY

- In this chapter, the effect of some characters of the myrmecochorous syndrome on the removal rates of diaspores by ants is analysed. The study of the morphology and dispersal of seeds of two closely-related species of *Corydalis* showed that large seeds of *C. cava* were more attractive to *Formica polyctena* ants and were removed more rapidly than small seeds of *C. bulbosa*.

- Field experiments included diaspores and juices of diaspore parts of *C. cava* and *Pulmonaria obscura* to test the effect of morphological and anatomical adaptations of diaspores and elaiosomes on diaspore attractiveness to *F. polyctena* ants. It was shown that chemical cues attracting ants are located in different parts of the diaspore: in the elaiosome (*C. cava, P. obscura*), in the coverage of the external layer of the fruit, and in the fruit collar (*P. obscura*).

- An investigation examined the effect of seed aggregation on seed removal rates by *F. polyctena* ants in a small-seeded diplochore *Chelidonium majus* and a large-seeded obligate myrmecochore *Viola odorata*. In *Ch. majus* the removal rates depended significantly on seed aggregation: seed removal was quicker from depots where seeds were closely aggregated. There were no significant differences in removal rates, obtained for depots with different distances between *V. odorata* seeds. It is concluded, that large single seeds of obligate myrmecochores have a higher chance of being repeatedly removed by ants after being dropped during transport than single seeds of small-seeded diplochores.

and seed removals at this microsite (Fig. 4.2). Tangents of the linear regression line slope for *L. fuliginosus* only and for summarised data for all three ant species were for visits: 0.3598 and 0.3422, respectively, and for removal: 0.2070 and 0.2015, respectively. For *M. rubra* and *L. nylanderi*, visiting and removal rates remain almost constant during 1.5 h of the experiment. For *M. rubra*, tangents of the linear regression line slope were -0.0192 for visits and 0.0066 for removal. For *L. nylanderi*, the tangents were 0.0028 and 0.0003, respectively. The number of refusals was considerably lower in both *L. fuliginosus* and *M. rubra* than in *L. nylanderi* (Fig. 4.2, B).

2. MICROSITE QUALITY FOR THE MYRMECOCHORE

One of the selective advantages of myrmecochory is that seeds escape predators, such as rodents and ground beetles from family Carabidae and Staphylinidae (see chapter 3, section 3.2). Furthermore, after fruit dehiscence, elaiosomes are normally exposed to a dry environment that may cause fast desiccation. Seeds with dry elaiosomes are unattractive to ants and are no longer properly dispersed (Levina, 1957) (Fig. 4.5). Ant attraction by elaiosomes therefore seems to be strongest just after fruit opening, and at this point the opportunity for seeds to be dispersed and escape predation is highest. It can be hypothesised that the microsite, which is able to provide the fastest seed removal by ants is advantageous for myrmecochores.

It is well known that diaspore removal rates by ants are determined by morphological characters of diaspores and by other characters of the myrmecochorous syndrome (Howe and Westley, 1988). Our results demonstrated that different ant species contribute differently to seed dispersal. Species with large colonies and large foraging territories, such as *F. polyctena* and *L. fuliginosus*, contribute much more to diaspore removal than species with a high number of small colonies and small-scale foraging territories (*M. rubra, L. nylanderi*). Predatory species (*F. polyctena*) more readily take seeds with elaiosomes than species that prefer a carbohydrate diet (*L. fuliginosus*). Size of ant workers considerably influences removal rates: species with large foragers removed large seeds of the myrmecochore *V. odorata* much faster than did small-sized ant species. In the species series *F. polyctena* – *L. fuliginosus* – *M. rubra* – *L. nylanderi*, in which body size decreased, contribution to seed removal decreased accordingly.

Since seeds have various removal rates by different ant species, this effects the total removal rate by an ant species complex in different microhabitats. Large seeds of *V. odorata* had the highest removal rate in microhabitats with *F. polyctena* (complex FP+MR+LN) dominating, this was lower in the territory of *L. fuliginosus* (complex LF+MR+LN) and very low in the territory of *M. rubra* (complex MR+LN). The entire content of the *V. odorata* fruit (10 – 35 seeds) might be removed between 0.5 – 1.5 h in the *F. polyctena* territory, between 1.0 – 3.0 h in the *L. fuliginosus* territory, and between 2.0 – 8.0 h in the *M. rubra* territory (Fig. 4.3). Since different microsites in the forest have different ant species complexes, the total removal rate of seeds depends on the particular species composition of the microsite. Thus, according to our data, if the «microsite quality» for a myrmecochore is preferably characterised by the removal rate, the most advantageous microsite for *V. odorata* is *F. polyctena* territory.

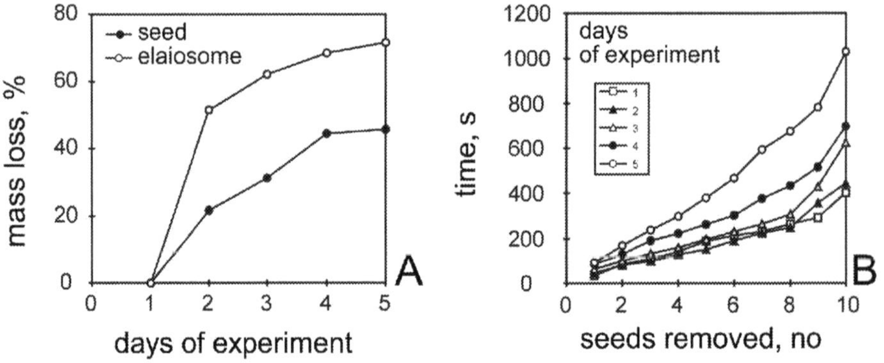

Fig. 4.5. Effect of elaiosome desiccation in seeds of *Viola odorata* on seed removal rates by *Formica polyctena* ants. A. Diaspore weight loss (black circles – seed; white circles – elaiosome). B. Seed removal dynamics for seeds at different stages of desiccation.

3. ELAIOSOME CONSUMPTION AND SEED REMOVAL

3.1 Duration of seed manipulation by ants

Ant workers do not take seeds immediately after recognition. They usually spend some time at the seed depot and test elaiosomes of several seeds. They manipulate the seeds for a while to be able to take them with their mandibles. Time of manipulation of seeds was considerably shorter in

M. rubra workers than in *F. polyctena* and *L. fuliginosus* workers (Table 4.1). At all microsites under consideration, manipulation time by *L. nylanderi* workers was considerably longer compared to other ant species studied (Table 4.1). In *F. polyctena* and, especially, in *L. fuliginosus* territories, *M. rubra* foragers spent much less time on a seed depot than they did on their own territory (Table 4.1, Fig. 4.6).

3.2 Number of seeds probed by ants prior to removal

If the seed number, probed by ants, is compared in two groups of foragers, (1) those that removed seeds and (2) those that refused removal, the highest difference is obtained for *L. fuliginosus* ants (Fig. 4.7, C). Workers of this species usually removed the very first seed probed. *M. rubra* workers probed a considerably higher number of seeds than *F. polyctena* and *L. fuliginosus* workers did. Workers of *F. polyctena, M. rubra*, and *L. fuliginosus* that finally refused to take seeds probed more than one seed more often than in cases where they finally did take a seed. Most individuals of *L. fuliginosus* that refused to take seeds probed more than three seeds (Fig. 4.7, C). *L. nylanderi* individuals usually probed more than three seeds, independently of subsequent action (removal or refusal) (Fig. 4.7, D).

Table 4.1. Time of ant manipulation with seeds (in seconds). Data are summarised for eight repetitions. Abbreviations: FP, *Formica polyctena*; LF, *Lasius fuliginosus*; LN, *Leptothorax nylanderi*; MR, *Myrmica rubra*; n, number of recordings; ref, manipulation with further refusal to remove a seed; rem, manipulation with further seed removal; sd, standard deviations; x_{ave}, average values.

Species		Microsite								
		FP territory			MR territory			LF territory		
		rem	ref	total	rem	ref	total	rem	ref	total
FP	x_{ave}	75.4	93.4	77.9	-	-	-	-	-	-
	sd	101.4	140.0	107.6	-	-	-	-	-	-
	n	809	128	937	-	-	-	-	-	-
MR	x_{ave}	48.5	52.6	51.2	111.8	91.3	101.3	19.2	24.5	20.9
	sd	52.4	69.4	63.7	111.4	93.9	103.2	28.5	24.2	27.2
	n	56	101	157	403	425	828	126	60	186
LF	x_{ave}	-	-	-	-	-	-	76.5	59.1	70.0
	sd	-	-	-	-	-	-	59.2	75.4	66.2
	n	-	-	-	-	-	-	1505	886	2391
LN	x_{ave}	1264.7	1085.0	1105.4	265.3	1033.5	948.2	238.6	415.8	398.7
	sd	1733.9	1319.0	1361.1	219.5	1084.6	1051.2	243.7	742.1	710.1
	n	10	80	90	3	24	27	5	47	52

3.3 Elaiosome consumption

Elaiosome consumption by ants over a long period of time without seed removal leads to a decrease of seed attractiveness to potential dispersal agents. Degree of the elaiosome consumption may be judged (1) from the duration of manipulation of ant individuals with seeds and (2) from the number of seeds probed by ants. In territories of *F. polyctena* and *L. fuliginosus*, ant workers of the dominating species consume elaiosomes a longer time before the seed removal than do *M. rubra* workers. Ants of the latter species consumed elaiosomes for a longer period of time, when foragers of larger ant species were absent from the seed depot.

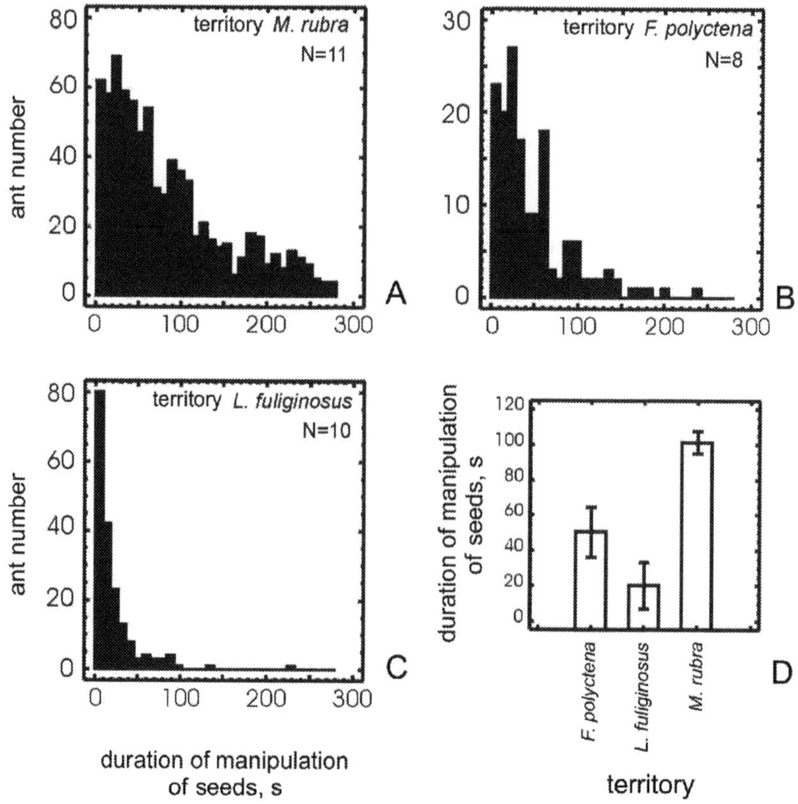

Fig. 4.6. Duration of manipulation of seeds by *Myrmica rubra* ants at the seed depots at different microsites. A – C. Frequency histograms. D. Means and standard deviations. N, number of experiments.

Workers of *F. polyctena, L. fuliginosus,* and *M. rubra* most frequently removed just the first seed probed. However, *M. rubra* ants probed a larger number of seeds than the two other species. In contrast to *F. polyctena* ants, *L. fuliginosus* and *M. rubra* individuals, after probing more than three seeds, usually refused seed removal. Workers of *L. nylanderi* consume elaiosome contents for a relatively long period of time without seed removal.

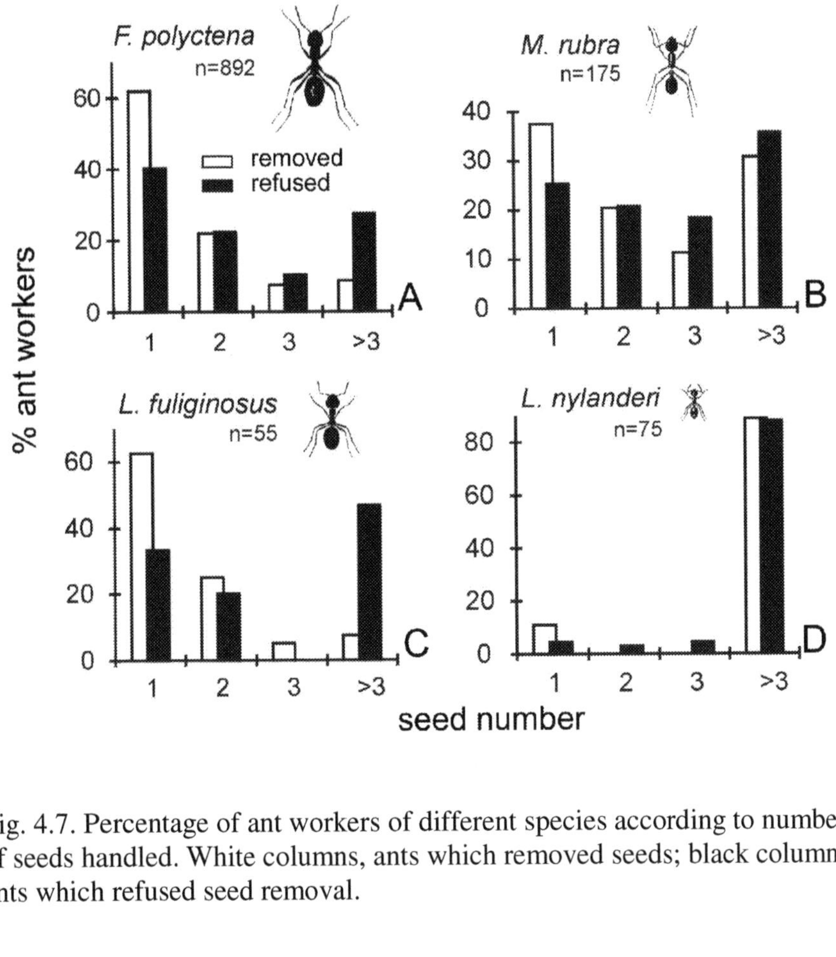

Fig. 4.7. Percentage of ant workers of different species according to number of seeds handled. White columns, ants which removed seeds; black columns, ants which refused seed removal.

3.4 Interactions between workers of different ant species at seed depots

Harvesting ant species that co-exist compete for food sources, and such a coexistence is supported by different levels of utilisation of accessible resources according to the type and size of food objects (Hansen, 1978).

Competition and similar division of foraging niches exist in carnivorous ant species and in those collecting aphid honey-dew. Such a competition may influence behaviour of ant individuals, especially of non-dominating species, at food sources such as seed depots.

In our experiments, *M. rubra* workers, under pressure by larger individuals of *F. polyctena* and *L. fuliginosus,* left seed depots much faster than when individuals of dominating species were absent. *F. polyctena* individuals frighten away foragers of *M. rubra*. Therefore, *M. rubra* ants could remove seeds only in the absence of *F. polyctena* workers. As a result of such a competition between ants, elaiosome consumption by *M. rubra* ants, without seed removal, was suppressed, which is advantageous for myrmecochorous plants. One may suggest that competition between two ant species may lead to a decrease in the number of visits to the seed depot and to a decrease of seed removal by one ant species. However, in our experiments such an effect was marginal.

4. EFFECT OF ANT RECRUITMENT BEHAVIOUR AND LEARNING ON SEED REMOVAL

4.1 Mark-recapture experiments

Features of ant foraging behaviour such as *recruitment* and *learning* may influence the dynamics of seed removal. In experiments with three ant species, *F. polyctena, M. rubra,* and *L. fuliginosus*, foragers taking seeds from the seed depot were marked. This method made it possible to compare arrival dynamics in two groups of individuals: marked («experienced») and unmarked («inexperienced»). Marked ants that arrived were considered «experienced», because they were able to find the depot once more after nest visiting or after dropping previous seeds during transport. Unmarked individuals arrived at the depot for the first time. Their arrival dynamics might only be influenced by the recruitment of experienced foragers.

Marked and unmarked *F. polyctena* ants arrived at the seed depot with a similar frequency during the experiment (Figs. 4.8, A; 4.9, A). In both groups, arrival rate of individuals increased, but the difference in the number of ant arrivals per unit of time was insignificant for both ant groups. Tangents of the linear regression line slope was 0.0527 for marked ants and 0.0620 for unmarked ants. There were also no differences in the duration of manipulation of seeds between marked and unmarked foragers of this species (Fig. 4.9, B).

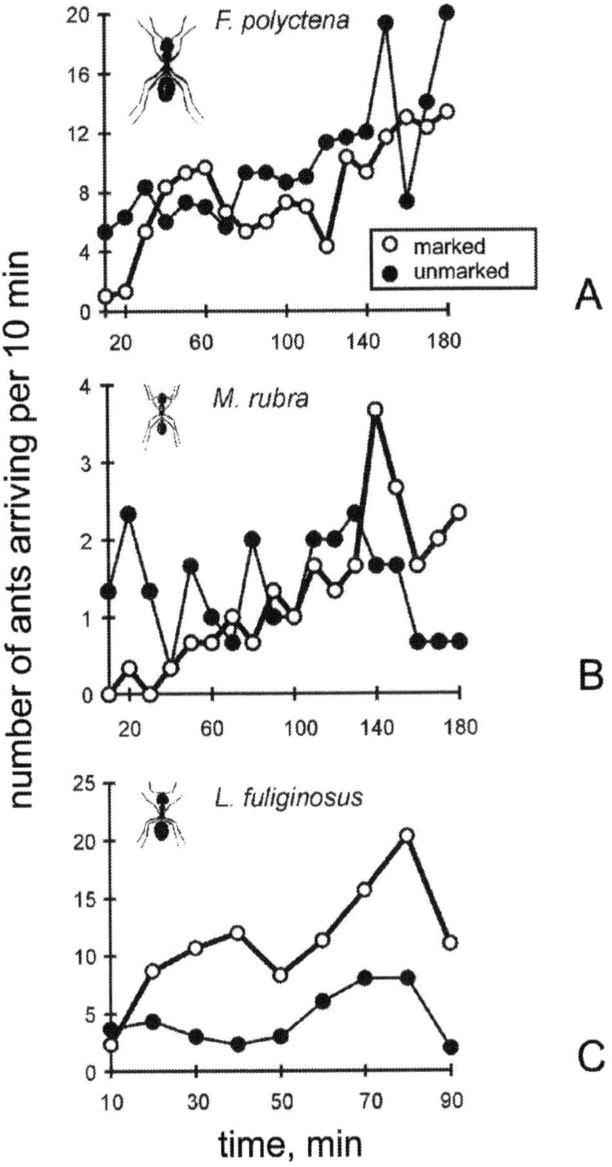

Fig. 4.8. Visits to the seed depot of marked and unmarked ant workers. Data obtained for three ant species: *Formica polyctena* (A, N=3 replicates, n=536 arrivals), *Myrmica rubra* (B, N=3, n=143), and *Lasius fuliginosus* (C, N=3, n=423).

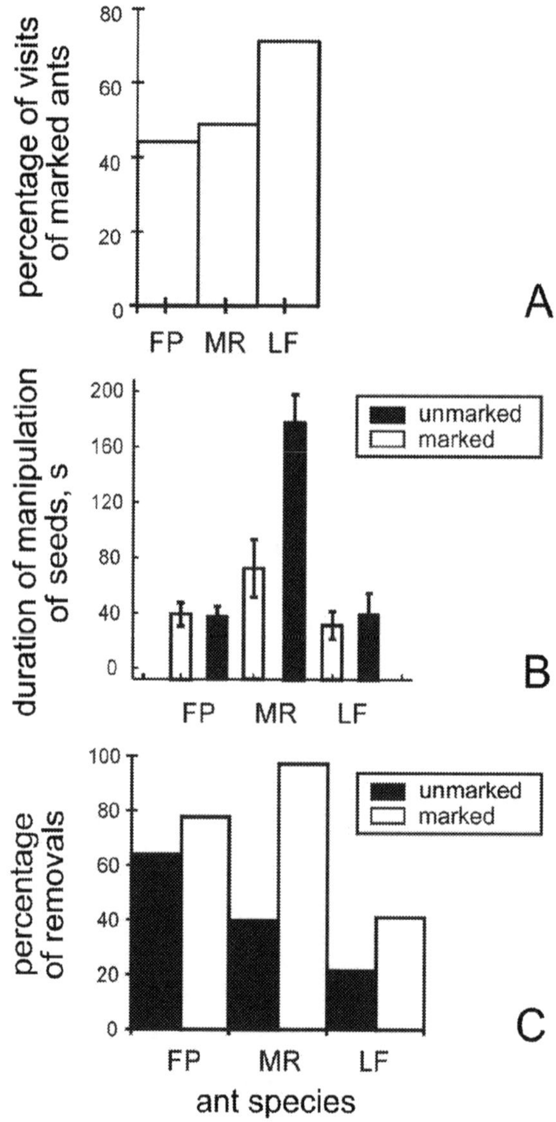

Fig. 4.9. Learning and recruitment effects on both visits and removal rates. A. Percentage of visits of marked ants over 3 h (A, B) or 1.5 h (C) of experiments. B. Duration of manipulation of seeds of marked and unmarked ants. C. Percentage of removal by marked and unmarked ants. FP, *Formica polyctena*; MR, *Myrmica rubra*; LF, *Lasius fuliginosus*.

In *M. rubra*, the number of individuals arriving was similar for both ant groups (Figs. 4.8, B; 4.9, A). However, the number of visits per unit of time was different for each group. Arrival rate of unmarked ants was just similar during the experiment (tangent of the linear regression line slope was -0.0014), whereas the number of marked ants per unit of time continuously increased (tangent of the linear regression line slope was 0.0155). Unmarked ants manipulated the seeds in the depot twice as much as marked ants (Fig. 4.9, B).

Marked *L. fuliginosus* visited the seed depot more often than unmarked ones (Figs. 4.8, C; 4.9, A). A difference in the visiting rates between the groups was insignificant. Visiting rate of marked individuals increased faster (tangent=0.1325) than unmarked ones (tangent=0.0300). Manipulation duration was similar in both groups (Fig. 4.9, B). Marked individuals of all three ant species (especially *M. rubra*) removed seeds more frequently than unmarked ones (Fig. 4.9, C).

4.2 Recruitment and learning in ants and seed removal rates

Learning is a change of behaviour under pressure of external factors. In ant workers, such a behavioural adaptation may lead to an increase in effectiveness of foraging (Johnson, 1991). Return of foragers to the food source may be a result of learning. In the experiments, described above, workers of all three species usually returned to food sources, although some differences in the number of individuals arriving per unit of time were registered.

Foragers of *L. fuliginosus* visited seed depots more often in the beginning of the depot exposure. After 70 – 80 min of exposure, the number of ants arriving rapidly decreased. This fact may be explained by the method of territory division by foragers of this species. In *L. fuliginosus*, each group of workers is responsible for one particular area of the entire territory (Dobrzanska, 1966). Therefore these ants could very quickly utilise seeds in the beginning of the depot exposure. When the majority of ant workers from this particular area of the territory are already involved in seed transport towards the nest, the number of foragers arriving to the seed depot for the first time decreases. In the case of *F. polyctena* and *M. rubra*, the number of ant visits to the seed depot increased during the exposure time.

Recruitment provides a chance for collective exploitation of the food source. Recruitment on food, described for many ant species (Chadab and Rettenmeyer, 1975; Horstmann, 1975; Cammaerts, 1978; Brian, 1983; Oliveira and Hölldobler, 1989), may considerably increase removal rates of seeds of myrmecochores. In *M. sabuleti*, recruitment was registered only in the case of a large food source (large piece or a large droplet of the extract of the victim) (Bisseau and Pasteels, 1994). Many small food objects, such as fruit flies, very rarely caused recruitment.

Diaspores of myrmecochores are small food objects. Therefore, we did not expect recruitment of foragers at the seed depot. However, an increase in the visiting rate of seed depots by unmarked ant workers may indicate that recruitment takes place in *F. polyctena* and *L. fuliginosus*. Stable visiting rate in *M. rubra* workers may indicate the absence of recruitment behaviour in this species.

It has been previously reported that searching behaviour of *Formica schaufussi* was more persistent for the carbohydrate-rich food than for the protein-rich food (Fourcassie and Traniello, 1993). Previous authors have considered that ants consider diaspores with elaiosomes as animal objects (Hughes et al., 1994). Return of foragers to the seed depot, recruitment behaviour, and persistence of searching behaviour, observed in our experiments, are behaviours supporting the hypotheses that ants indeed do perceive seeds with elaiosomes as animal objects.

Prior to seed removal, both «experienced» and «inexperienced» foragers of *F. polyctena* and *L. fuliginosus* manipulated seeds for a relatively short time (30 – 50 s), only consuming some amount of tissues in elaiosomes. «Inexperienced» *M. rubra* manipulated seeds twice as long as «experienced» workers did. During recurring visits, they spent much less time consuming elaiosomes. Therefore, it may be suggested that due to learning, a large number of seeds remain with undamaged elaiosomes after ant probing.

Since «experienced» ant workers removed seeds more frequently than «inexperienced» ones, it is evident that learning and recruitment influence seed dispersal. Learning and recruitment contribute to faster diaspore removal, and may prevent thus elaiosome desiccation. In this case, probability of seed predation by rodents and beetles decreases. Presumably, these effects play an important role for obligate myrmecochores, whose seeds remain in groups after fruit opening.

Hence, results of field experiments described in this chapter, show that an ant species complex of a particular microsite may be an important factor influencing diaspore removal rates of myrmecochorous plants.

5. SUMMARY

- Diaspores of myrmecochorous plants are usually removed by a variety of ant species. Ant species contribute differently to the removal of elaiosome-bearing seeds.

- Microhabitats on the deciduous forest floor differ in the composition of ant species. These differences may influence the dispersal of myrmecochorous plants.
- This chapter presents the results of a study of the effect of ant species composition on removal rates of *Viola odorata* seeds by foragers of *Formica polyctena* (FP), *Lasius fuliginosus* (LF), *Myrmica rubra* (MR), and *Leptothorax nylanderi* (LN) at three microhabitats: *F. polyctena* territory with FP, MR, LN; *L. fuliginosus* territory with LF, MR, LN; and *M. rubra* territory with MR, LN. It is shown that, at each territory, the main contribution to visiting rates and to seed removal rates was made by the most dominant ant species.
- Ant competition for seeds results in a decrease in the consumption of elaiosomes without seed dispersal by *M. rubra* workers, and this could be advantageous for plants. Competition between two ant species may lead to a decrease of both visiting and removal rates by one species. However, this effect was statistically insignificant in our experiments.
- Learning (*M. rubra*) or recruitment (*F. polyctena*) took place during seed removal. Experienced ants removed seeds with higher frequencies than inexperienced ones regardless of ant species. It is concluded that learning and recruitment strongly effect the seed removal rates. This gives the myrmecochore a better chance of being quickly removed and thus prevents the drying out of the elaiosomes. Additionally, it reduces exposure to potential predators.
- Ant species composition strongly influences the removal rates of seeds and potentially the fate of the seeds of a myrmecochore. *F. polyctena* territory, provides the fastest removal of the *V. odorata* fruit content, and is presumably the most advantageous for the myrmecochore.

Chapter 5

Diaspore transporting by ants

As previously mentioned, myrmecochory is true mutualism (Beattie, 1985; Handel and Beattie, 1990): non-seed-eating ants consume energetically valuable elaiosomes and transport seed for some distance from the parent plant. This has been registered in representatives of many groups of plants and ants (Beattie and Lyons, 1975; Culver, Beattie, 1980; Nesom, 1981; Clay, 1983; Mesler and Lu, 1983; Wein and Pickett, 1989; Bond and Stock, 1989; Higashi et al., 1989; Baroni-Urbani and Nielsen, 1990; Hughes and Westoby, 1990; Pacini, 1990). Ant workers collect diaspores with elaiosomes and carry them to their nests. Some of the diaspores reach the nests and are concentrated there, whereas others are dropped during transport. Dropped diaspores fill up the soil seed bank and provide territory occupation by plant offspring. Additionally, dropped diaspores may be secondarily removed by other ant individuals or even by workers of other ant species and transported further towards ant nests. Dropped portions of seeds and portions of secondarily found and removed seeds are factors which may aid in the evaluation of the role ants play in diaspore dispersal for distance and in understanding the general strategy of diaspore dispersal. One may suggest that these factors depend on the plant species and, in the first instance, on the diaspore and elaiosome dimensions, as well as on the ant species and average size of ant individuals.

In this chapter, we estimate seed dropping during transport by *Formica polyctena* ants in two species of obligate (*Asarum europaeum, Viola hirta*) and three species of facultative myrmecochores (*Chelidonium majus, V. matutina, V. mirabilis*) (for a detailed method description see chapter 11, section 4.2.5). Since this ant species has large foraging territories compared to *Myrmica rubra* and *Leptothorax nylanderi* (see chapter 11, section 3), one may suggest a considerable influence of ant workers on the dispersal distance. In the present chapter, we (1) analyse methods of diaspore transport by ants, (2) discuss seed dropping rates during transport, (3) quantify size classes of two groups of foraging individuals: individuals that removed seeds from the depot, and individuals that removed and relocated seeds for some distance (3 m) from the seed depot, and (4) by using a computer simulation, analyse effects of seed producing capacity and seed dropping rate during transport on dispersal distance of diaspores.

1. METHODS OF DISPORE TRANSPORTING

Methods of a load transport are different among ant species. *F. polyctena* workers usually transport other ant individuals with the front part of the body facing forward, whereas *Myrmica rubra* workers do just the opposite and carry the body with the hind part facing forward (Wilson, 1971). Harvesting ants use various methods for seed transport (Bernard, 1971; Johnson, 1991). Method of seed holding during transport is determined by the relationship between the morphological feature and size of diaspores and by the size of an ant worker. Since ant species of the deciduous forest differ in size, some interspecific differences in transporting methods exist.

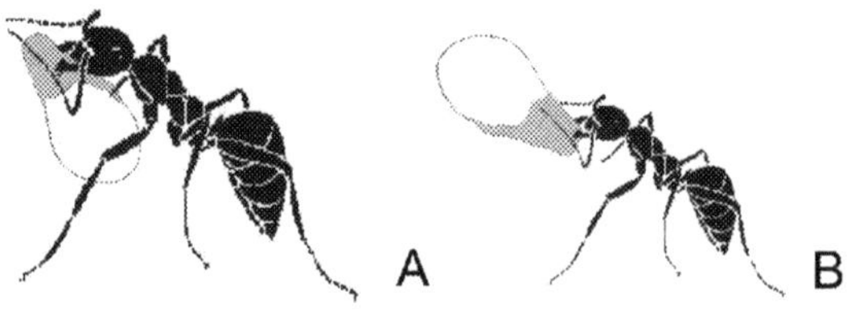

Fig. 5.1. Methods of holding of small (A) and large (B) seeds by ants *Formica polyctena* during transport.

To remove the diaspore from the depot, an ant selects the region on the diaspore, the size of which best corresponds to an optimal span of the ant mandibles. Observations of the transport of *V. odorata* seeds have shown that large ant species (*F. polyctena*) tend to transport these seeds with the narrow side forward (Figs. 5.1, A; 5.2). If the maximal size of the diaspore is larger than ant clearance during walking (distance between the ventral side of the worker thorax and substrate surface), ants tend to transport seeds with the broad side forward (Fig. 5.1, B). The majority of *M. rubra* foragers, small individuals of *F. polyctena*, and approximately one half of *L. fuliginosus* individuals use the latter method for the transport of *V. odorata* seeds (Fig. 5.2). Since seeds of this plant are rather heavy for small ants *Leptothorax nylanderi*, workers of this species usually apply reverse walking and drag the seed along the substrate.

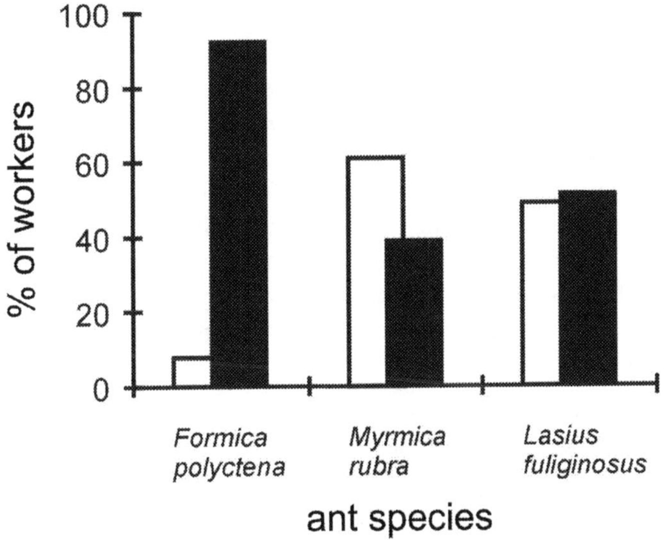

Fig. 5.2. Methods of holding *Viola odorata* seeds by ants of *Formica polyctena, Myrmica rubra*, and *Lasius fuliginosus* during transport. White columns, seed transport with the narrow end forward; black columns, seed transport with the broad end forward.

In *F. polyctena*, transporting methods may also depend on the size and morphological features of diaspores. In the case of large seeds of *A. europaeum, V. hirta,* and *V. mirabilis*, ants take them by elaiosomes. Small seeds of diplochores (*Ch. majus, V. matutina*) are taken by seed bodies. Foragers of *F. polyctena* transport middle-sized seeds (*V. mirabilis*) with the narrow part facing forward, and large seeds (*V. hirta*) with the broad part facing forward. Small seeds (*Ch. majus*) are transported by either one of these methods. In the case of large seeds of *A. europaeum*, small ant individuals may apply reverse walking and drag them along the substrate. *L. nylanderi* always do this with *V. odorata* seeds. Furthermore, large seeds of *A. europaeum* may be transported by two ant individuals simultaneously. Presumably, transporting methods may influence diaspore dropping during transport. However, this hypothesis was not tested in our experiments.

2. DIASPORE DROPPING DURING TRANSPORT

Dropping rates of seeds during transport by ants differ in various plant species (Fig. 5.3). In mark-recapture experiments, more than 35% of seeds of *A. europaeum* were relocated by *F. polyctena* foragers to a distance of 3 m from the seed depot. About 30% of dropped seeds reached 3 m due to other ant individuals, which have secondarily taken dropped seeds from the forest litter. In the series of species *A. europaeum* – *V. hirta* – *V. mirabilis* – *Ch. majus,* the proportion of dropped seeds increased. None of the *V. matutina* seeds reached 3 m. The majority of seeds of this diplochore was dropped at a distance of 10 – 30 cm from the seed depot. In the series of species *A. europaeum* – *V. hirta* – *V. mirabilis*, the number of seeds, picked up by other ant individuals, increased considerably. Secondary removal of dropped seeds from the forest litter was very rarely observed in small-seeded diplochores (*Ch. majus, V. matutina*).

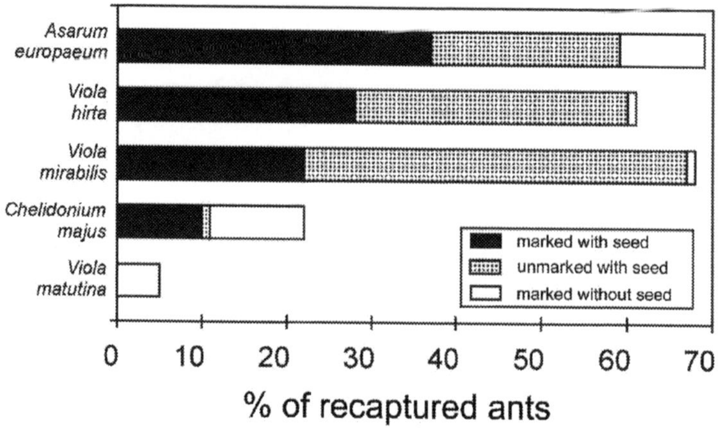

Fig. 5.3. Results of mark-recapture experiments. Percentage of recaptured ants that reached 3 m distance from the initial point (seed depot). Black bars, marked ants that transported marked seeds for a distance of 3 m; grey bars, unmarked ants that picked up dropped marked seeds (repeated removal); white bars, marked ants that dropped marked seeds. In experiments with each plant species, 100 marked ants were used.

Thus, knowing that the dropping rate during transport accounts for 8 – 10% of large-seeded myrmecochores and 20 – 100% of small-seeded diplochores, one may assume that the main portion of small seeds do not reach the nest, usually located at a distance of 3 – 20 m from the plant. Most of the seeds roll into the forest litter or soil in the territory of the ant colony.

An increase in seed dropping rate during transport in the series of plant species *A. europaeum* – *V. hirta* – *V. mirabilis* – *Ch. majus* – *V. matutina* correlates well with a decrease of seed removal rates (Gorb and Gorb, 1995b). Since seeds of obligate myrmecochores (*A. europaeum, V. hirta*) and large seeds of some diplochores (*V. mirabilis*) are rarely dropped by ant workers, they can be transported over a large distance by ants. If these seeds are dropped, they can be secondarily detected and removed from the forest litter by other ant individuals. In the series of species *A. europaeum* – *V. hirta* – *V. mirabilis*, an increase of seed dropping rate is compensated for by the relatively high secondary removal rate. Probably, the high chance for repeated removals is also given by the large dimensions of the seed and elaiosome. Small seeds of diplochores (*Ch. majus, V. matutina*) are more often dropped and more seldom repeatedly removed, because they usually roll much deeper into the forest litter or soil.

Thus, two alternative strategies of seed dispersal for distance are revealed in the plant species studied. (1) Obligate myrmecochores *A. europaeum* and *V. hirta* produce a small number of large seeds having a high removal rate and a low dropping rate during transport by ants. (2) Diplochores *V. matutina* and *Ch. majus* produce numerous small seeds having a low removal rate and a high dropping rate. Diplochores with large seeds (*V. mirabilis*) are intermediate between these extreme variants. In nature, disadvantages of the myrmecochorous stage of dispersal is compensated in diplochores by the autochorous dispersal stage. Diplochorous plants use ants as additional dispersal agents (Ulbrich, 1928; Levina, 1957; Berg, 1966; Beattie and Lyons, 1975; Beattie and Culver, 1979; Westoby and Rice, 1981; Ohkawara and Higashi, 1994).

3. DEPENDENCE OF DIASPORE DROPPING ON THE ANT WORKER SIZE

Sizes of ant workers removing *V. hirta* seeds did not differ from those that transported seeds over a distance of 3 m (Fig. 5.4, B, D). Data analysis shows that there was no difference between the size of ant individuals that removed and transported *V. hirta* seeds (Fig. 5.4, E). However, in the case of *A. europaeum* seeds, such differences were revealed. On the frequency histogram of size classes of ants transporting seeds, the maximum peak is shifted considerably towards the larger sizes (Fig. 5.4, A, C). Thus, small ant individuals dropped *A. europaeum* seeds more often than larger ants (Fig. 5.4, E).

Frequency histograms of the head width of ant workers, which removed seeds, were shifted towards larger ant sizes for large-seeded plant species (*A.

europaeum and *V. hirta*) and towards smaller ant sizes for small-sized plant species (*Ch. majus* and *V. matutina*), compared to the data on ant foragers randomly collected from the forest litter in the territory of the same ant colony (see Fig. 2.4, chapter 2). Thus, data of the field experiments let us suggest that dispersal distance of diaspores depends, at least, on two processes: (1) seed removal and (2) seed transport by ants. Each process has some constraints for different plant species, and these constraints are dependent on the ant workers' dimensions.

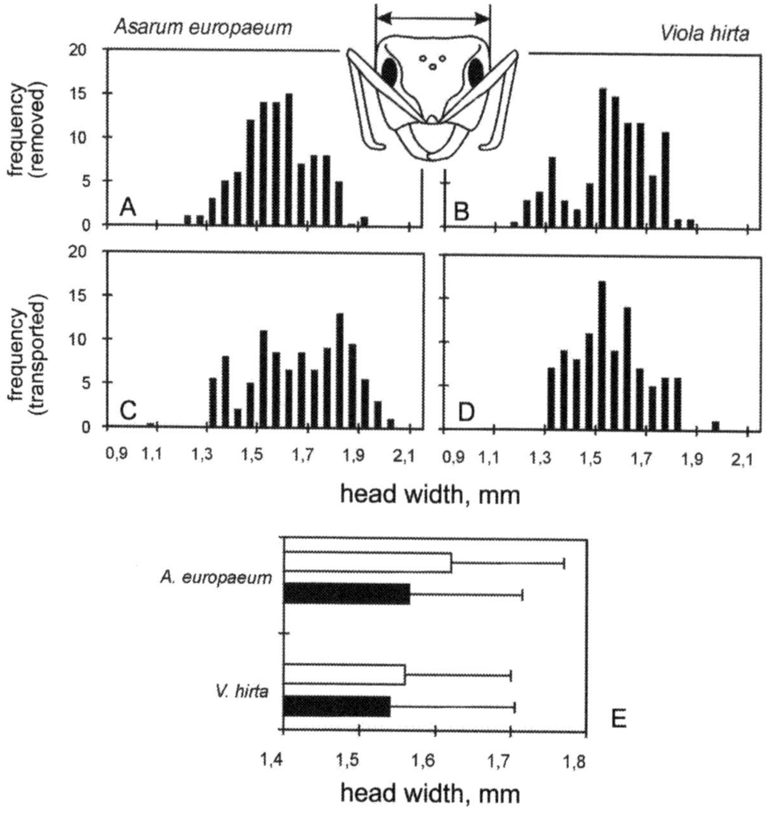

Fig. 5.4. Size classes of *Formica polyctena* workers. A – D. Frequency histograms of the head width measured in workers that removed seeds (A, B) and transported seeds (C, D) of *A. europaeum* (A, C) and *V. hirta* (B, D) over a distance of 3 m. E. Average values and standard deviations of head width measured in workers that removed (black bars) and transported (white bars) seeds.

4. EFFECT OF DIASPORE DROPPING ON DISPERSAL DISTANCE

4.1 Complex of factors used in the computer model

To estimate the influence of both seed producing capacity and seed dropping rate during transport by ants on the seed dispersal distance, a computer simulation was applied. The model system consists of the parent plant (with co-ordinates x=0, y=0) producing N seeds with the dropping portion, L, during transport by ants over a distance of 3 m. Co-ordinates of each seed were determined relative to the parent plant location as follows. Ants transport seeds in a randomly selected direction. In reality, *F. polyctena* workers move along foraging routes. Since we did not analyse the influence of the ant's route pattern on the spatial seed distribution, the situation was somewhat simplified to concentrate on the effect of seed dropping rate on the dispersal distance. The single distance S, along which L percent of seeds were dropped, was set to 3 m, because this single distance was previously used by authors in the experimental set-up (see chapter 11, section 4.2.5). Within each single distance (S), an amount of dropped seeds (L) was randomly distributed.

By varying values of N and L, different variants of plant strategies were tested. As a result, a sample of dispersal distances was obtained for different plant strategies. Thus, the simulation was based on two factors: portion of seeds dropped along the ant route (L), and general seed producing capacity of the plant (N). Both variables were previously estimated for real plant species studied. To evaluate differences in distribution of dispersal distances for species with different morphological adaptations of seeds, the L variable was set from 1% to 64%. To evaluate differences in distribution of dispersal distances for species with different seed producing capacity, N was set from 30 to 5000 seeds.

4.2 Computer model

Computer simulation generated seed patterns based on different values of L and N variables (Fig. 5.5, A, B). The patterns were used to calculate maximal dispersal distances, which can be reached by seeds at a certain combination of L and N values. Some patterns were generated according to data obtained in field experiments (Fig. 5.5, C, D).

Figs. 5.5, A, B demonstrate influence of the dropping rate on dispersal distance with similar seed producing capacity of the plant. Figs. 5, C, D show different strategies of plants to occupy the same territory. Facultative myrmecochores (Fig. 5.5, C) produce a large number of small seeds, having a high dropping rate during transport by ants, whereas obligate myrmecochores (Fig. 5.5, D) produce a few large seeds, having a low dropping rate.

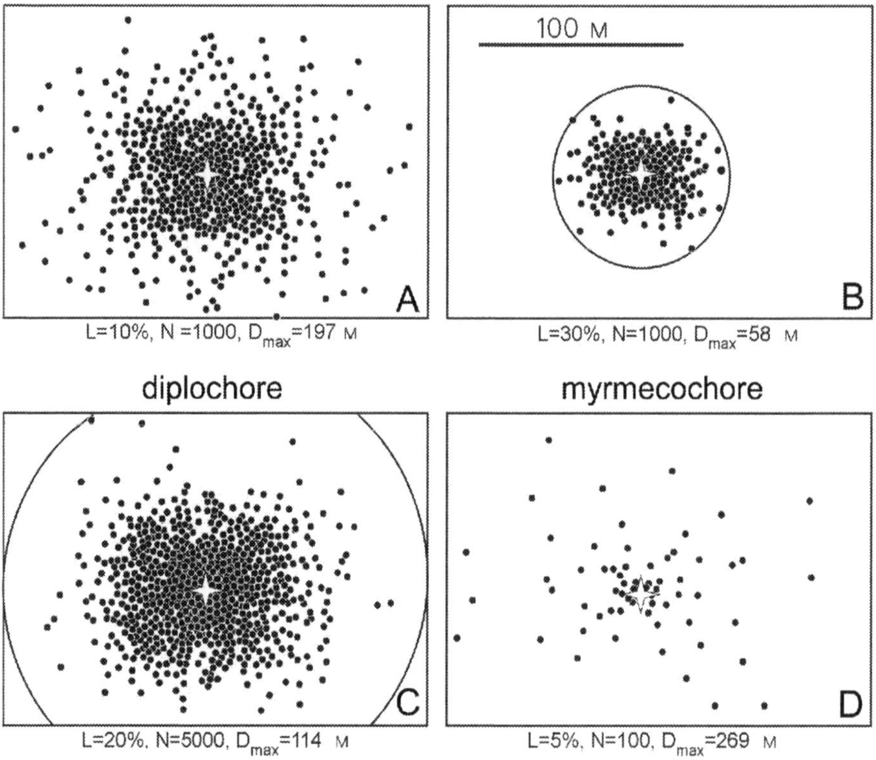

Fig. 5.5. Computer-simulated patterns of seed dispersal distances for plant species with different values of seed producing capacity (N) and seed dropping rates (L). A and B show the effect of seed dropping on the dispersal distance of seeds. C and D show different ant-related dispersal strategies of plants. Star, a parent plant; D_{max}, maximal possible distance which seeds may reach under given N and L; point, a seed. Information below each diagram indicates the values set to variables in each simulation. Scale bar: 100 m.

These strategies can be better compared on charts (Fig. 5.6) generated on the basis of the analysis of the computer model. Both an increase of the seed producing capacity and a decrease of the seed dropping rate lead to an increase of maximal dispersal distance of seeds (Fig. 5.7). Although the present simulation is based on a limited number of specific factors, it reflects well the various strategies of diaspore dispersal for distance in myrmecochorous plants.

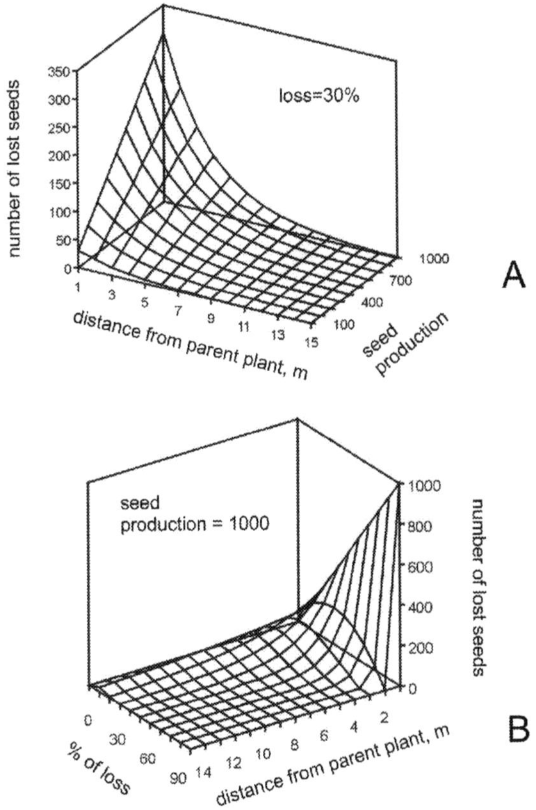

Fig. 5.6. Computer-simulated diagrams showing the number of seeds, dropped at some distance from the parent plant as a function of seed producing capacity under a constant dropping rate (A) and a function of the dropping rate under a constant seed producing capacity (B). $n=(N \cdot L \cdot (1-L/100)^{D/S-1})/100$, where D is distance from parent plant (in m); L – seed dropping rate within S (in %); n – number of seeds dropped at some distance from the parent plant; N – number of seeds produced by the parent plant; S – single distance along which L % of seeds are dropped (3 m).

Thus, data from literature, field experiments, and computer simulation let us conclude that, at each stage of diaspore dispersal (recognition, removal, transport), there is a kind of filter determining the number of diaspores passing through it. These filters present the difference in diaspore removal rates and the difference in diaspore dropping rates during transport and are different for various plant species. The filters depend on both diaspore and elaiosome dimensions, as well as on the size of ant foragers and species-specific features of ant behaviour.

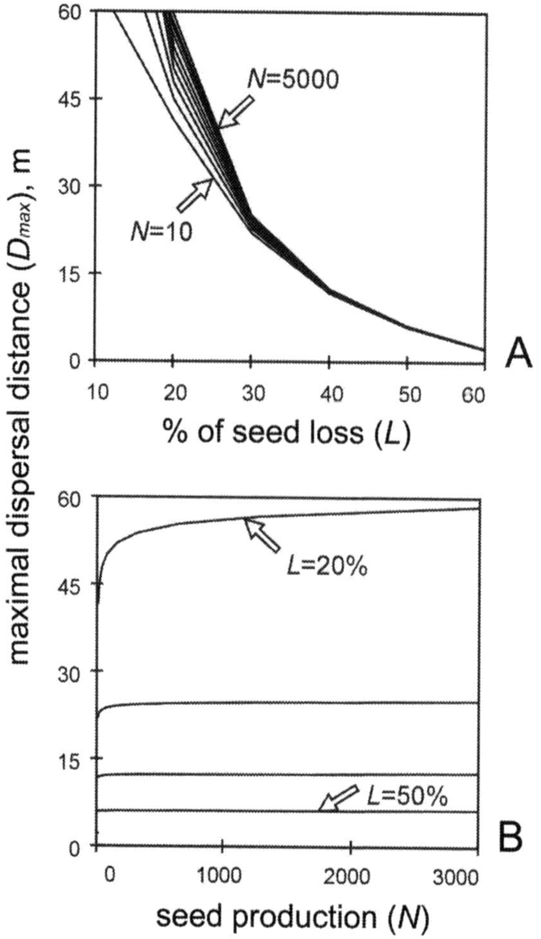

Fig. 5.7. Maximal dispersal distance reached by seeds (D_{max}), as a function of seed dropping rate (A) or seed producing capacity (B). $D_{max}=-[(S \cdot lgN)/lg(1-L/100)]$, where N is number of seeds produced by the parent plant; L – seed dropping rate within the single distance (in %); S – single distance along which L % of seeds are dropped (3 m).

5. SUMMARY

- In a deciduous forest, foraging ants collect elaiosome-bearing seeds and carry them to their nests. Some of the seeds reach the nest and are concentrated there. Others may be dropped by ants during transport. The dropped seeds enter the soil seed pool. However, they might be repeatedly removed by other ant individuals and again carried in the direction of the nest.

- Rates of seed dropping and repeated removals are important parameters to evaluate the effect of ant workers on dispersal distance of seeds. The chapter presents the results of mark-recapture experiments carried out to evaluate the dropping rates of seeds of five plant species (*Asarum europaeum, Viola hirta, V. matutina, V. mirabilis, Chelidonium majus*) during transport by *Formica polyctena* ants. It is shown, that the rate of seed dropping depends on the size of seeds and of elaiosomes, both of which vary among plant species, and on the size of ant workers.

- In the species series *A. europaeum* – *V. hirta* – *V. mirabilis* – *Ch. majus* – *V. matutina*, the dropping rate increases. Across species, dropping rates of seeds are negatively correlated with the rate at which ants remove them from the depot. Small workers drop seeds of *A. europaeum* more often than large workers do, while seeds of *V. hirta* are dropped by ants of different size classes with the same frequency.

- The number of seeds which reach nests depends on dropping rates: seeds with lower dropping rates have a higher chance of being deposited in nests. Diplochores usually produce many small seeds, which are characterised by low removal rates and high dropping rates during transport. Obligate myrmecochores produce rather few large seeds, which have high removal rates and low dropping rates.

- To analyse the significance of seed dropping for dispersal distance of seeds, a computer simulation based on two factors, seed number produced by a plant and dropping rate of seeds, is proposed. It is calculated that an increase in seed producing capacity and a decrease in seed dropping rate result in an increase of maximal dispersal distance.

Chapter 6

Seed flow in ant territories

One of the main questions in studies of seed dispersal by ants is the fate of diaspores, removed by ants (Beattie and Lyons, 1975; Culver and Beattie, 1978; 1980; Beattie, 1983; Kjellsson, 1985; Higashi et al., 1989; Ohkawara and Higashi, 1994; Gorb and Gorb, 1995a; b; Gorb, 1998). Previous authors have shown that diaspores with elaiosomes are actively collected by ants and transported to their nests. There, ants usually gnaw away elaiosomes and throw out the diaspore bodies, which retain their ability for germination (Sernander, 1906; Culver and Beattie, 1983). Moreover, mechanical and probably chemical treating of diaspores by ants may influence seed germinating ability. Our experiments showed that intact seeds of *Viola odorata* taken directly from capsules germinated considerably less well than seeds manipulated by ants (Appendix, Fig. IX, C). It was previously found for two violet species that even after three years, survival rates of plants germinated from seeds taken from ant nests were higher than in the control group. In *Corydalis aurea*, not survival rates, but seed producing capacity of ant-manipulated plants was twice as high as in controls (Handel and Beattie, 1990).

Diaspore bodies are usually concentrated in the vicinity of ant nests (Appendix, Fig. VIII). Ants may strongly influence seed flow within their territories and diaspore concentration in their nests. Consequences of such an influence on the seedling distribution and plant associations have been studied in several publications (Malozemova and Koruma, 1970; Beattie and Culver, 1977; Radchenko et al., 1979; Gorb and Gorb, 1995c). It has been shown that the peak of plant species diversity is found at a distance of 1.5 m from nest mounds of ants of the genus *Formica*. This effect is explained by soil peculiarities of these microhabitats (transition from nest substrate to normal soil). It is also suggested that ant workers deposit waste material, including diaspore bodies of myrmecochorous plants, at a similar distance from the nest. It has been previously reported that the location of ant colony *garbage dumps* varies even within a genus of ants. For example, *Formica cunicularia* deposit waste into special chambers located deep in the soil. *F. cinerea* workers concentrate waste in the vicinity of nest entrances. Representatives of the subgenus *Formica* s. str. and *F. execta* transport waste to the border of their territory and leave them there (Dlussky, 1967).

The spectrum of plant species in various microhabitats depends on the seed material present in the soil. It may be hypothesised that distribution of seeds of various plant species in the soil in territories of different ant species depends on the biology of the specific ant species. Distribution of seedlings and mature plants on nest mounds of different ant species has been studied by several authors (Malozemova and Koruma, 1973; King, 1977a; b; c; Beattie and Culver, 1977; 1981; Radchenko et al., 1979; Culver and Beattie, 1983; Woodell and King, 1991). However, there is lack of information on the seed material of nest mounds. In this chapter, seed flow in the deciduous forest is analysed using data obtained (1) from direct observations of seed transport on the foraging routes of *F. polyctena* (for a detailed method description see chapter 11, section 4.2.6) and (2) from the study of soil seed pools of ant nest mounds and randomly selected control areas on the foraging territory of the ant colony (for methods see chapter 11, section 4.3). Two groups of plant species were compared according to their adaptability to the ant-dispersal: myrmecochores and non-myrmecochores.

1. DIRECT OBSERVATIONS OF THE SEED FLOW

Field observations showed that, in May, at the peak of fruiting of early-spring ephemeral myrmecochores, intensity of seed flow is at an average of 600 – 750 seeds/h·m (Fig. 6.1). Such a high intensity remains for at least three weeks. Since periods of fruiting in different myrmecochores are not exactly the same, but rather vary to some extent (Gorb, 1998) (see Fig. 1.10, chapter 1), the occurrence of specific plant species among seed material transported to the nest was also rather different (Fig. 6.1).

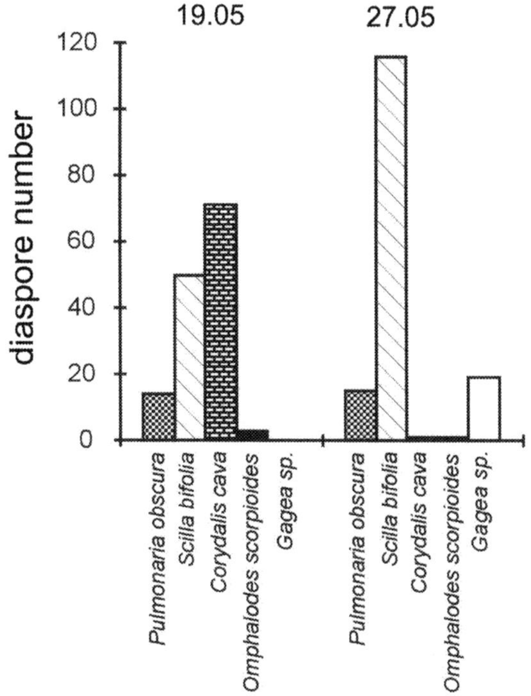

Fig. 6.1. Seed flow of diaspores of early-spring ephemeral myrmecochores on foraging routes of *Formica polyctena* ants. Data were collected during 1 h on the ant route 20 cm wide at a distance of 4 m from the nest (t=25 °C).

2. SOIL SEED MATERIAL

2.1 Herbaceous species

Data on seed material of 17 plant species that were found in the soil samples are presented in Table 6.1. For herbaceous species, the total seed number and the number of plant species were similar in samples from ant mounds and in control samples (total seed number: nest – 613.1, control – 948.7; species number: nest – 11.0, control – 11.5), although the total number of seeds was somewhat higher in control samples.

To characterise microsites according to the soil seed pool, principal component analysis (PCA) was applied. Results of PCA are illustrated in Figs. 6.2 – 6.4. To evaluate the role of plant species differing in methods of diaspore dispersal, three samples were analysed: the first sample included all herbaceous species (17), the second one – myrmecochores only (9), the third one – non-myrmecochorous species drawn to mounds of *F. polyctena* nests (6).

Table 6.1. Seed material from soil samples of the foot of the nest mounds of *Formica polyctena* (n) and randomly chosen areas of ant territories (f). Abbreviations: x(n) and x(f), means for samples from nest mounds and from territories, respectively. Myrmecochores are marked with an asterisk.

Plant species	Number of diaspores in samples										
	n1	f1	n3	f3	n4	f4	n5	f5	f6	x(n)	x(f)
**Corydalis cava*	275	407	143	357	262	1311	52	736	145	183.0	591.2
**Corydalis bulbosa*	124	123	11	43	1	4	70	8	37	51.5	43.0
**Viola odorata*	110	122	38	63	34	73	11	9	110	48.3	75.4
**Lamium maculatum*	28	15	21	41	11	19	4	0	19	16.0	18.8
**Ballota nigra*	38	69	0	12	0	0	0	0	40	9.5	24.2
**Asarum europaeum*	6	0	0	0	0	1	0	0	0	1.5	0.2
**Omphalodes scorpioides*	0	0	5	0	0	2	0	0	0	1.3	0.4
**Pulmonaria obscura*	0	0	1	0	2	3	0	0	0	0.8	0.6
**Scilla bifolia*	0	0	2	10	0	8	0	4	5	0.5	5.4
Sambucus nigra	738	32	11	123	22	14	39	4	110	202.5	56.6
Galium aparine	5	1	313	33	59	91	69	3	3	111.5	26.2
Robinia pseudoacacia	1	2	57	17	118	35	266	301	0	110.5	71.0
Anthriscus sylvestris	79	10	313	6	4	10	3	0	3	99.8	5.8
Fallopia convolvulus	134	21	0	2	3	0	0	1	25	34.3	9.8
Stellaria holostea	27	29	45	41	25	26	7	39	18	26.0	30.6
Alliaria petiolata	1	1	94	0	10	1	0	0	38	26.3	8.0
Geum urbanum	6	10	0	3	2	0	0	0	0	2.0	2.6
Aegopodium podagraria	1	3	0	7	1	2	0	0	11	0.5	4.6
Glechoma hirsuta	0	2	1	11	0	0	0	0	0	0.3	2.6
TOTAL: diaspores	1573	847	1055	769	557	1600	521	1105	564	926.1	977.1
species	15	15	14	15	14	15	9	9	13	13.0	13.4

For the first sample, the first principal component (PC) accounts for 73.8% of variance, the second one – 13.4%. Areas, corresponding to samples from the foot of the nest mounds and to control samples, do not overlap (Fig. 6.2). Control samples form a compact cluster in the right lower corner of the field of the first two principal components. This may be evidence for the high similarity of control soil samples. Samples from nest mounds were considerably scattered, and their cluster is diagonally lengthened. Thus, the second PC contributes much more strongly to the sample division than the first PC. Myrmecochores have the strongest positive (*Corydalis cava, Viola odorata, C. bulbosa*) and negative (*Asarum europaeum, Pulmonaria obscura, Omphalodes scorpioides*) contributions to the division of microsites according to the first PC. Non-myrmecochores, which tend to grow on the ant mounds (*Galium aparine, Anthriscus sylvestris*) have the main positive contribution according to the second PC.

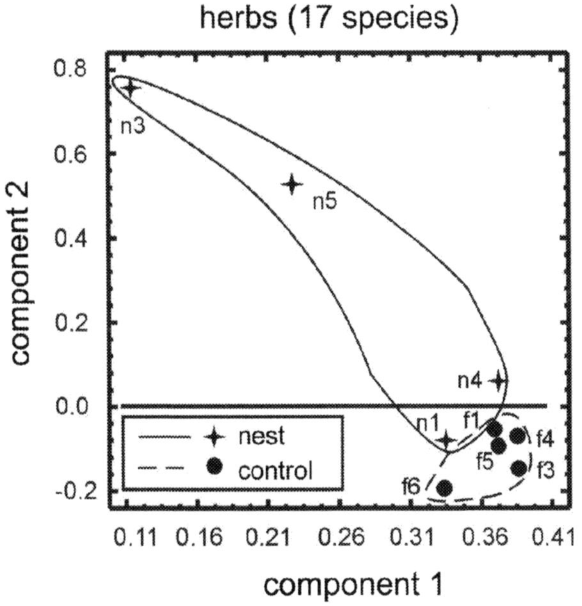

Fig. 6.2. Principal component analysis of soil samples. Plots of the first two component weights for seeds of 17 plant species from nine soil samples. The first component accounts for 73.8% of variance, and the second one – 13.4%. n1 – n5, samples from ant nests; f1 – f6, control samples.

2.2 Myrmecochores

In the soil samples, diaspores of nine myrmecochorous plants were found (Table 6.1). All species were detected in samples from ant mounds and in control samples. In spite of the prevalence of large-seeded myrmecochores in the seed flow of foraging routes directed towards the nest, their prevalence in the samples from ant mounds was not detected. Furthermore, seeds of myrmecochores usually prevailed in control samples. Above all, this was found with large-seeded myrmecochores, such as *Viola odorata, Pulmonaria obscura, Corydalis cava, Lamium maculatum, Scilla bifolia*, as well as with some small-seeded ones (*Ballota nigra, C. bulbosa*).

Interestingly, PCA analysis of nine species of myrmecochores did not divide the microsites studied (Fig. 6.3). The first PC accounts for 87.7% of variance, the second one – 8.3%. Although control samples form a compact cluster in the right lower corner of the field of the first two principal components, areas corresponding to samples from ant mounds and to control samples overlap almost up to 90%. This may indicate similarity of soil seed samples from both types of microsites. Only two samples from ant mounds (n1, n5) do not fit a common cluster. A cluster of samples from ant mounds is diagonally lengthened. In this case, the first PC contributes more to division of samples than the second one. The main positive contribution to the sample division according to the first PC was made by *Corydalis cava, Viola odorata,* and *C. bulbosa*, according to the second PC – *V. odorata* and *C. bulbosa*.

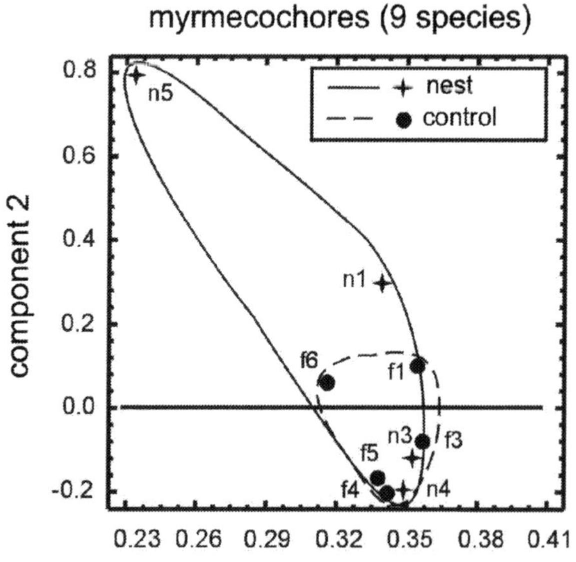

Fig. 6.3. Principal component analysis of soil samples. Plots of the first two component weights for seeds of nine myrmecochorous plant species from nine soil samples. The first component accounts for 87.7% of variance, and the second one – 8.3%. n1 – n5, samples from ant nests; f1 – f6, control samples.

2.3 Non-myrmecochores

In the PCA analysis, plant species having diaspores without elaiosomes were considered as non-myrmecochores. Diaspores of eight non-myrmecochorous plant species were found in soil samples (Table 6.1). All species were detected in samples from ant mounds and in control samples.

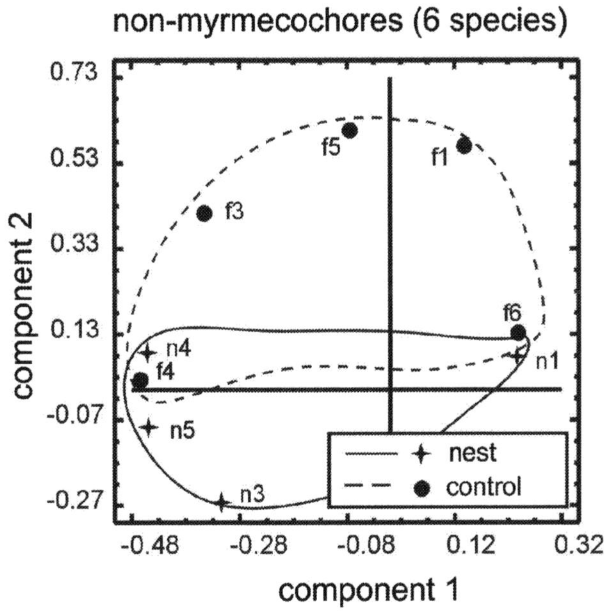

Fig. 6.4. Principal component analysis of soil samples. Plots of the first two component weights for seeds of six non-myrmecochorous plant species from nine soil samples. The first component accounts for 47.6% of variance, and the second one – 26.3%. n1 – n5, samples from ant nests; f1 – f6, control samples.

In contrast to myrmecochores, which were divided into large- and small-seeded ones, more appropriate division for non-myrmecochores was based on the tendency of their diaspores to be concentrated in ant nests. The first group has this tendency and consists of *Anthriscus sylvestris, Galium aparine, Glechoma hirsuta, Alliaria petiolata, Stellaria holostea,* and *Fallopia convolvulus*. The second group does not have this tendency and consists of *Geum urbanum* and *Aegopodium podagraria*. Diaspores of non-myrmecochores were rarely observed as being transported by ants. In spite of the absence of the diaspores in the seed flow directed to the ant nests, the diaspores of non-myrmecochores prevail in soil samples from ant mounds (*Anthriscus sylvestris, Galium aparine, Alliaria petiolata, F. convolvulus*).

For PCA analysis, only six species belonging to the first group were selected. The first PC accounts for only 47.6% of variance, the second PC – 26.3%. Samples from ant mounds and control samples do not form compact clusters (Fig. 6.4). Although clusters are strongly scattered along both principal components, they do not considerably overlap. Each cluster occupies the lower (samples from ant mounds) or upper (control samples) half of the field. Here, not the first, but rather the second PC mainly contributes to the division of samples. *Fallopia convolvulus* and *Alliaria petiolata* contribute most positively to the sample division along the first PC. *Galium aparine* and *Stellaria holostea* contribute most negatively to the sample division along the first PC. *F. convolvulus*, *S. holostea* (positively), *Alliaria petiolata,* and *Anthriscus sylvestris* (negatively) contribute to the sample division along the second PC.

3. EFFECT OF ANT WORKERS ON THE VEGETATION

In the literature, many examples are described about how ants directly effect vegetation. Ant workers may cover seedlings growing on the ant mound with soil or bite plant roots and thus kill plants (Woodel and King, 1991). There are also indirect effects, such as change of plant species abundance on ant mounds because of changed physical and chemical properties of the soil (Greenfeld, 1941; Czervinski et al., 1971; Malozemova and Koruma, 1973; King, 1977a; b; c; Petal, 1978; Davidson and Morton, 1981; Beattie and Culver, 1983; Culver and Beattie, 1983). A number of non-myrmecochores may concentrate in the surroundings of *F. polyctena* mounds (Appendix, Fig. VIII) mainly due to the high content of nutrients in the soil of ant mounds (Dlussky, 1967; Gorb et al., 1997; Sindarovskaya et al., 1997). The spectrum of plant species in various habitats may also depend on the seed material deposited in soil. Local distribution of such a soil seed bank may be influenced by ant species composition of the microsite.

4. SEED FLOW ON THE TERRITORY OF THE *FORMICA POLYCTENA* COLONY

4.1 Diaspore concentration in ant nests

Soil seed material is different in both types of microsites (the foot of the nest mounds of *F. polyctena* and control samples from the forest). This difference reflects the result of the behaviour of ant workers of this species and dispersal strategies of plants. Let us make a course estimation of the number of diaspores of myrmecochores that reach the nest. By using a direct count of diaspores in the field on ant foraging routes, it was found that, in late May, approximately 1200 – 1500 diaspores of myrmecochores reach the nest per hour (overall width of large routes at the perimeter of the ant mound averages approximately 2 m). In 10 h, in daytime, 14400 – 18000 diaspores will reach the nest. In 20 days of the fruiting period of early-spring ephemeral species of myrmecochores, approximately 288000 – 360000 diaspores will be deposited in the nest. Additionally, ant workers collect some diaspores during the entire season, not only during the peak of the fruiting of myrmecochores. If *F. polyctena* ants, after consumption of the elaiosomes, just throw away diaspore bodies from the nest and deposit them in the vicinity of the foot of the nest mound, the diaspore density in the upper soil layer would reach 1000 – 1500 diaspores/dm^2 at the end of just one season (for an average ant mound diameter of 1 m and mound foot width of 25 – 30 cm). Since soil samples mentioned in previous sections were taken from 3 – 4 year old ant nests, one might expect a diaspore density of 3000 – 6000 diaspores/dm^2. In reality, diaspore density of myrmecochores in processed soil samples was 20 times lower than expected. This contradiction between estimated data and results of studies of soil samples may be explained by the phenomenon of the secondary relocation of diaspore bodies after elaiosome consumption in *F. polyctena* ants (see chapter 7).

4.2 Flow of diaspores of myrmecochores

Diaspores of some species of myrmecochores, such as *Asarum europaeum* and *Pulmonaria obscura*, do not germinate in nests of *F. polyctena*. It was not clear, whether these species have other advantages from myrmecochory in addition to the dispersal for distance. Even an advantage from dispersal for distance can be neglected for plant individuals in terms of growing in the vicinity of the ant nest, because almost all seeds will reach the nest (Gorb and Gorb, 1999a). This is especially true for species having diaspores with a low dropping rate during transport by ants. However, our data show that diaspores of myrmecochores do not concentrate in the nest, in spite of a considerable diaspore flow towards nests. Diaspores without elaiosomes may be transported by *F. polyctena* ant workers to the territory border together with other waste materials (see chapter 7). This would explain why the density of diaspores of myrmecochores was lower in the seed pool of ant mounds than in the soil seed pool of the territory (control samples).

Interestingly, on the smooth surface of diaspore bodies of the large-seeded myrmecochores, some prominent structures are present. They function as handles enabling diaspore bodies transport even after elaiosome removal. Dimensions of these handles correspond to the mandible span of ant workers and are presumably optimised for diaspore body holding by an ant individual (Fig. 6.5).

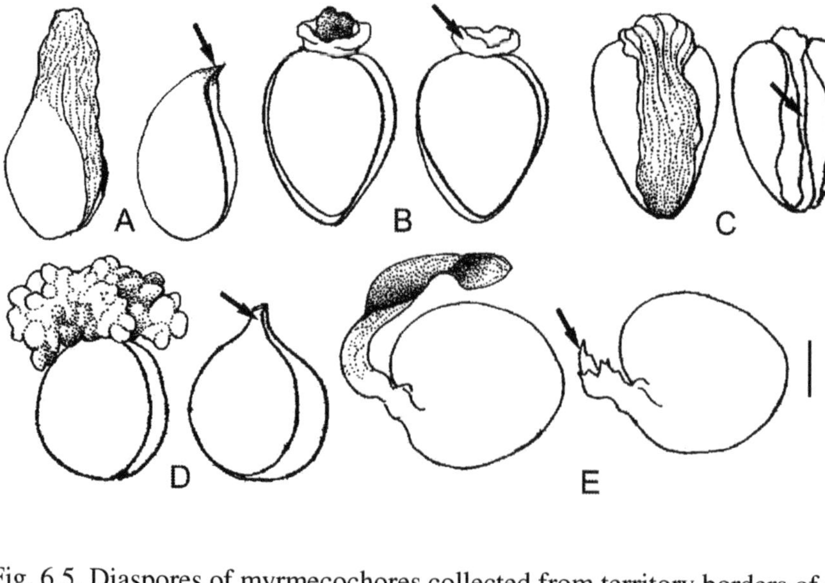

Fig. 6.5. Diaspores of myrmecochores collected from territory borders of the *Formica polyctena* ant (left), and intact diaspores (right). A. *Viola odorata.* B. *Pulmonaria obscura.* C. *Asarum europaeum.* D. *Scilla sibirica.* E. *Corydalis cava.* Arrows indicate structures allowing the holding of a diaspore by ants after elaiosome removal. Scale bar: 1 mm.

Thus, diaspore bodies of myrmecochores not only leave *F. polyctena* nests, but they are transported by ant workers a few meters in the direction opposite to the nest. It is reasonable to assume that a large portion of seeds is dropped during such a secondary transport and enriches the soil seed pool on the colony territory. Consequently, diaspores flow in both directions: the primary or centripetal one (directed to the nest), and the secondary or centrifugal one (directed away from the nest to the territory border). Both transporting processes increase diaspore dispersal. Having two diaspore flows, even plants growing in the vicinity of *F. polyctena* nests have the chance for diaspore dispersal, because their seeds will be secondarily redispersed within the colony territory.

4.3 Flow of diaspores of non-myrmecochores

A slightly different picture is observed in diaspores of non-myrmecochores. Diaspores of many species (*Fallopia convolvulus, Alliaria petiolata, Glechoma hirsuta, Anthriscus sylvestris, Stellaria holostea, Galium aparine*), whose dimensions correspond to those of building materials used for the mound construction, may be removed from the forest litter and transported, by ant workers, to their nests. It seems that features of nest-building behaviour of *F. polyctena* workers promote dispersal of some diaspores lacking elaiosomes (see also chapter 10). If these diaspores reach the nest, they concentrate in the ant mound and are not subject to secondary transport from the nest, as in the case of elaiosome-bearing seeds. These diaspores have the advantage of germination in the nutrient-enriched substrate of the mound (Gorb and Gorb, unpublished data), if ants do not kill them by biting their roots. However, diaspores of non-myrmecochorous plants, grown on the ant mound, are doomed to remain in the nest without dispersal. Therefore, seeds of non-myrmecochores gradually concentrate in the mound substrate (this chapter, section 2.3). Many other non-myrmecochorous species were reported as tending to concentrate on mounds of *Formica* spp. ants (*Urtica dioica, Rubus saxatilis, Rosa canina, Rubus idaeus, Stellaria* sp.) (Dlussky, 1967). Possibly, gradual concentration of the seed material in the nest sooner or later lead to the nest being overgrown with vegetation. This effects nest shading and may result in the colony leaving the nest.

According to the nutrient-enrichness theory (Handel, 1978a; Culver and Beattie, 1978; 1980; 1983; Beattie, 1983; Handel and Beattie, 1990; Gibson, 1993a; b), the main advantage of myrmecochory is occupation of ant nests by some myrmecochores, because of the nutrient-enriched nest substrate. Other authors do not consider such an occupation as a selective advantage of diaspore dispersal by ants and give their view that only diaspore relocation over distance from the parent plant is an advantage supported by the evolutionary process (Horvitz and Schemske, 1986a; Rice and Westoby, 1986; Andersen, 1988; Bond and Stock, 1989; Bond et al., 1991; Hughes, 1991). Dispersal-for-distance hypothesis states that, in most cases of myrmecochory, it is not the occupation of nutrient-enriched habitats but rather the decrease of competition from the parent plant and other seedlings.

Our data on an interaction between *F. polyctena* ants and the variety of plant species show an inverse effect than that predicted by the nutrient-enrichness hypothesis for myrmecochorous plants. Diaspores of myrmecochores do not concentrate in nests of this ant species. This fact contradicts the main claim of the nutrient-enrichness hypothesis. For non-myrmecochores, whose diaspores concentrate in ant nests, mound substrate provides some advantages to the growth and development of seedlings (Gorb et al., 1997). At the same time such a concentration undoubtedly leads to an increase of the competition between plants that grow on the mound substrate. Since myrmecochory is a very complex interaction, which involves not only different species of plants and ants, but rather communities of plants and ants, one may judge its advantages only by considering an interaction between a pair of species in a specific environment. We think, however, that dispersal over distance, at least for plants of a temperate deciduous forest, is presumably the main driving mechanism of the evolution of this method of diaspore dispersal.

Mechanisms considered in the present chapter may influence plant species composition in various microhabitats of a deciduous forest and participate in plant guilds' formation (Beattie and Culver, 1981) (Appendix, Fig. IX, B).

5. SUMMARY

- Ants can influence the direction of seed flow within the territory of an ant colony and the concentration of diaspores in nests. In this chapter, the results of direct observations of seed transport on foraging routes and the analysis of soil seed material from the nest mounds of *Formica polyctena* ants were presented.

- It is shown that both the total number of diaspores, and the number of plant species in soil samples from ant nests (the foot of the nest mounds) and territories (control) are not significantly different. The total number of diaspores was somewhat higher in control samples.

- Diaspores of non-myrmecochorous species dominate in nest samples. Although myrmecochores prevail in the diaspore flow on foraging routes directed to nests, they do not dominate in soil samples from nest mounds.

- After removal of elaiosomes, diaspores of myrmecochores are transported by ant foragers to territory borders (secondary relocation phenomenon). To evaluate how plant species with different diaspore dispersal adaptations contribute to the soil seed bank, a principal component analysis was used. Three data sets were analysed: seeds of all herbaceous species (17 species), of myrmecochores (9), and of non-myrmecochores tending to grow on the ant mounds (6).

- Diaspores of non-myrmecochores with size classes corresponding to those of nest building material may be transported to nests due to the nest-building behaviour of ants. Diaspores of such plants, originating from diaspores which were brought into nests by ants, remain in the nests and are not relocated to territory borders. Mechanisms described may influence the plant species composition in various microsites of deciduous forest and formation of guilds of myrmecochores.

Chapter 7

Secondary relocation of diaspores from ant nests

In previous chapters, it was shown that ants may influence diaspore flow and promote diaspore redistribution within their territories. How does such ant activity result in the vegetation pattern of various microsites in the territory of the ant colony? Knowing that *Formica polyctena* ants transport diaspore bodies to the territory borders, it is important to clarify the role of the borders for myrmecochorous plants. For this reason, it is interesting to compare soil seed pools and vegetation of the borders with those of the rest of the ant territory and their nests.

In the present chapter, we report a case study of secondary relocation of diaspores in territories of two ant species (*F. polyctena* and *Lasius fuliginosus*). Distribution of mature plants, seedlings, and soil seed pools were studied in territories and their borders in two ant species having different ecology and behaviour (for a detailed method description see chapter 11, sections 4.2.7 and 4.3). Experiments were carried out in the area of the forest with three nests of *F. polyctena* and one nest of *L. fuliginosus*, where territories of both species have a common border. Since the border location changes with the colony growth and exists, as such, only for some period of time, the new and old borders were studied.

Ant species studied deposit waste materials at different places. *L. fuliginosus* store them in the vicinity of the nest, whereas *F. polyctena* transport them to territory borders. Such a difference in the location of garbage dumps suggests that behaviour of these ant species has a different effect on the diaspore dispersal and plant distribution in the deciduous forest. In this chapter, the following three main questions are resolved: (1) Are there differences in distribution of mature plants and seedlings at microsites studied? (2) Does seed material from ant nests differ from that of territory borders and from the soil seed pools of territories of both species? (3) Are there differences in seed material from similar microsites between ant species studied?

1. COMPOSITION OF MATURE PLANTS AND SEEDLINGS IN VARIOUS MICROSITES

1.1 Mature plants

There was no considerable difference in density of mature plants among four microsites studied. In territories of *F. polyctena* and *L. fuliginosus*, the density was approximately 27.2 and 65.6 plants/m^2, respectively; on the old and new borders – 60.8 and 32.0 plants/m^2, respectively. Data on mature plants of 14 plant species found altogether in four microsites are given in Table 7.1. There was no significant difference between territories and borders in the plant species number. Within experimental squares of 0.25 × 0.25 m, 5 – 6 plant species were found in territories of *F. polyctena* and *L. fuliginosus*, whereas on borders there were 6 – 8 species. Analysis of the occurrence of mature myrmecochores showed that only one species (*Viola odorata*) was present at all microsites, *Asarum europaeum* – only on the new border, *Pulmonaria obscura* – only on the old border, and *Ballota nigra* – only on *L. fuliginosus* territory.

1.2 Seedlings

Seedlings (Fig. 7.1) are irregularly distributed among various microsites (Table 7.2). On the old border, 1177.6 seedlings/m^2 were counted, on the new border – 547.2, in *L. fuliginosus* territory – 235.2, in *F. polyctena* territory – 342.4. In total, seedlings of 21 plant species were detected: 13 – on the new border, 15 – on the old border, 13 – in *L. fuliginosus* territory, 17 – in *F. polyctena* territory.

There was considerable difference in the seedling density between various microsites. In each square, seedlings of 5 – 6 plant species were registered. At each type of microsite, 5 species of myrmecochores were found (only *Pulmonaria obscura* was absent in the territory of *L. fuliginosus*), that constitute 28.5% of all registered plant species. Seedlings of some species were very rare (1 – 2 individuals). Amounts of seedlings of two species (*Asarum europaeum* and *Viola odorata*) were much higher on the borders of territories of both species than within territories (Table 7.2). For *A. europaeum*, an average of 220.8 seedlings/m^2 were counted on the new border, 699.2 – on the old border, 8.0 – in *L. fuliginosus* territory and 27.2 – in *F. polyctena* territory. For *V. odorata*, 107.2 seedlings/m^2 were counted on the new border, 201.6 – on the old border, 43.2 – on the *L. fuliginosus* territory and 70.4 – on the *F. polyctena* territory. To compare microsites according to the data on seedling amounts of various plant species, principal component analysis (PCA) was applied (Fig. 7.2). The first component accounts for 46.0% of variance, the second one – 24.0%. Regions corresponding to both territories overlap at least by 90%. Regions corresponding to both borders also considerably overlap.

Table 7.1. Numbers of mature plants counted within experimental squares (25×25 cm, n=40) in *Formica polyctena* territory (FP), in *Lasius fuliginosus* territory (LF), and on borders between territories of these two ant species. Abbreviations: ave, average number of plants in the square (n=10); fr, number of squares, in which the plant species was found; mx, maximal number of plants per square; *, myrmecochores.

Plant species	New border			Old border			LF territory			FP territory		
	ave	mx	fr	ave	mx	fr	ave	mx	fr	ave	mx	fr
* *Asarum europaeum*	0.2	2	1	-	-	-	-	-	-	-	-	-
* *Viola odorata*	0.4	2	3	0.1	1	1	0.4	2	2	0.2	1	2
* *Pulmonaria obscura*	-	-	-	0.1	1	1	-	-	-	-	-	-
* *Ballota nigra*	-	-	-	-	-	-	0.5	3	3	-	-	-
Anthriscus sylvestris	0.1	1	1	-	-	-	-	-	-	-	-	-
Fallopia convolvulus	0.9	3	4	0.1	1	1	0.6	5	2	-	-	-
Geum urbanum	0.1	1	1	0.1	1	1	-	-	-	0.1	1	1
Rubus caesius	-	-	-	-	-	-	0.1	1	1	-	-	-
Glechoma hirsuta	-	-	-	1.7	4	7	1.9	8	6	1.2	5	3
Prunella vulgaris	-	-	-	0.3	2	2	-	-	-	-	-	-
Stellaria holostea	-	-	-	0.6	4	3	0.6	2	3	0.1	1	1
Geranium robertianum	-	-	-	-	-	-	-	-	-	0.1	1	1
Veronica chamaedrys	-	-	-	0.8	8	1	-	-	-	-	-	-
Euonymus verrucosa	0.2	2	1	-	-	-	-	-	-	-	-	-

Data on seedlings from both borders form a compact cluster in the right upper part of the graph. Only three experimental squares from the new border have small values of the second principal component (PC). Regions corresponding to both territories and those corresponding to borders do not overlap considerably. Both principal components contribute to the division of groups of data corresponding to «borders» and «territories». Two myrmecochores (*Asarum europaeum, Viola odorata*) and one non-myrmecochore (*Anthriscus sylvestris*) have the highest positive load on the first PC. *Asarum europaeum* is the only myrmecochore with a considerable positive load in the second PC. Two species, myrmecochore *Ballota nigra* and non-myrmecochore *Anthriscus sylvestris*, have high negative load in the second PC.

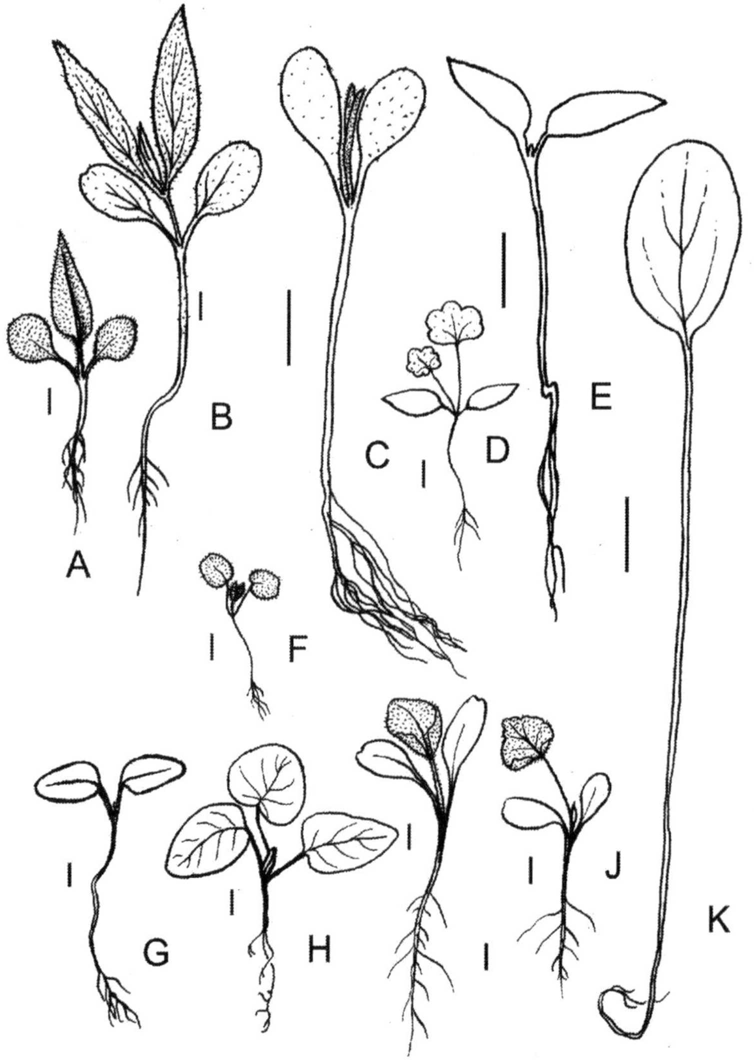

Fig. 7.1. Seedlings and juveniles of myrmecochores. A. *Pulmonaria obscura*. B, C. *Omphalodes scorpioides*. D, E. *Chelidonium majus*. F. *Lamium maculatum*. G, H. *Asarum europaeum*. I. *Viola odorata*. J. *V. mirabilis*. K. *Corydalis cava*. Scale bar: 5 mm.

Table 7.2. Number of seedlings in experimental squares (25×25 cm, n=40) in *Formica polyctena* territory (FP), in *Lasius fuliginosus* territory (LF) and on borders between territories of the ant species. Abbreviations: ave, average number of plants per square (n=10); fr, number of squares, in which seedlings of the plant species were found; mx, maximal number of plants per square; †, trees; *, myrmecochores.

Plant species	New border			Old border			LF territory			FP territory		
	ave	mx	fr	ave	mx	fr	ave	mx	fr	ave	mx	fr
* *Asarum europaeum*	13.8	31	10	43.7	80	10	0.5	4	2	1.7	4	8
* *Viola odorata*	6.7	15	9	12.6	21	10	2.7	17	6	4.4	10	9
* *Pulmonaria obscura*	1.1	3	7	1.0	4	4	-	-	-	1.7	5	7
* *Ballota nigra*	2.1	4	8	1.8	6	8	1.9	5	6	1.4	3	6
* *Chelidonium majus*	0.7	6	2	0.8	5	3	0.6	2	5	1.0	6	3
Anthriscus sylvestris	2.2	8	6	6.7	15	7	5.2	10	10	4.3	15	8
Fallopia convolvulus	0.1	1	1	-	-	-	0.1	1	1	-	-	-
Alliaria petiolata	0.6	1	5	0.1	1	1	0.9	4	4	0.9	7	2
Geum urbanum	0.3	2	2	0.2	2	1	0.1	1	1	0.7	4	3
Rubus caesius	-	-	-	-	-	-	0.2	1	1	-	-	-
Glechoma hirsuta	0.4	4	1	1.9	7	5	0.3	2	2	1.1	3	6
Atriplex sp.	0.1	1	1	-	-	-	-	-	-	0.1	1	1
Stellaria holostea	-	-	-	0.3	2	2	0.2	1	2	-	-	-
Moehringia trinervia	-	-	-	0.1	1	1	-	-	-	0.2	1	2
Geranium robertianum	-	-	-	-	-	-	-	-	-	0.1	1	1
Veronica chamaedrys	-	-	-	2.2	10	3	-	-	-	0.4	3	2
†*Fraxinus excelsior*	0.1	1	1	-	-	-	0.3	1	3	0.1	1	1
†*Acer platanoides*	-	-	-	0.2	1	1	-	-	-	0.3	1	3
†*Ulmus laevis*	-	-	-	-	-	-	-	-	-	0.1	1	1
†*Quercus robur*	-	-	-	0.1	1	1	-	-	-	-	-	-
Undetermined	6.0	18	9	1.9	7	5	1.7	6	8	2.4	7	8

1.3 Effect of ants on the distribution of plants in the forest

Effects of ant workers on plants, growing on ant mounds are widely described in the literature (Lebedev, 1949; Kostrin, 1951; Grimasky, 1960; Dlussky, 1967; Malozemova, 1970; Davidson and Morton, 1981a; b; Culver and Beattie, 1983; Gorb et al., 1996), but the vegetation on the territory borders has not been previously studied. According to the hypothesis of the nutrient-enriched habitats (see chapter 6, section 4.3), diaspores of myrmecochores have an advantage by getting into the nutrient rich substrates of ant nests.

Territory borders between colonies of *F. polyctena* and *L. fuliginosus* are a kind of garbage dump for waste storage by *F. polyctena* workers. These dumps contain organic remains (dead ant workers, pieces of insect cuticle, and diaspores of myrmecochores with removed elaiosomes) (Appendix, Fig.

X). Therefore, these microsites may be considered as nutrient-enriched habitats. However, according to our data, complexes of mature plants on the borders between territories and in the territories of both ant species do not differ. Probably, this is because of the annual dynamics of borders.

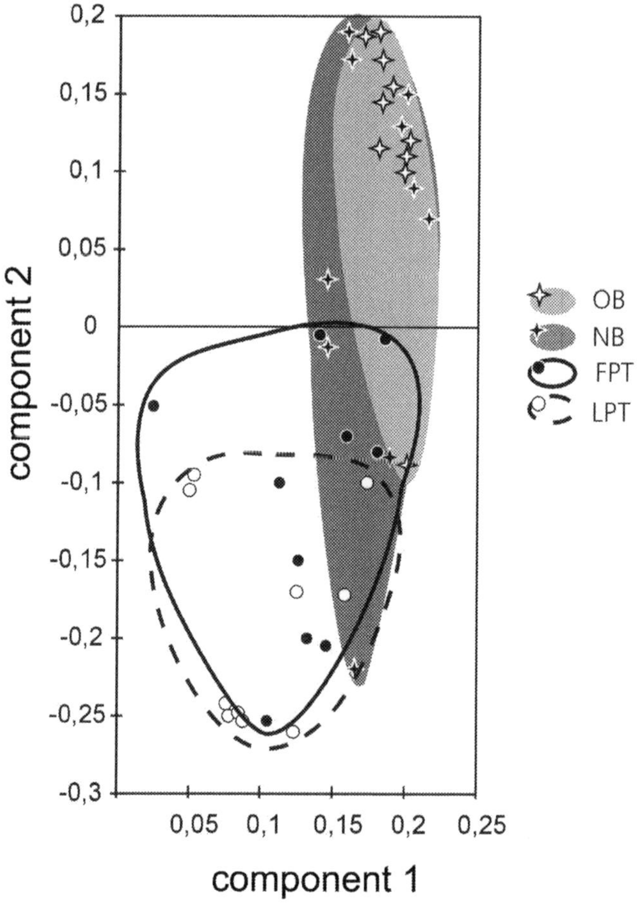

Fig. 7.2. Principal component analysis of seedling samples. First two component weights for seedlings of 21 plant species from 40 sample plots from six microsites. The first component accounts for 61.5% of variance, and the second one – 24.3%. FPT, *Formica polyctena* territory; LFT, *Lasius fuliginosus* territory; NB, new border; OB, old border.

Waste material was rather fresh on both borders. The fact that myrmecochorous plants require several years for development could explain why mature plant composition on borders is the same as elsewhere in ant territories. Presumably, later, after diaspore germination and rotting of waste material, the plant composition of this microsite will become different from that of the surrounding microsites. However, by this time it will probably no longer be the territory border.

Since the amount of seedlings of large-seeded myrmecochores (*Asarum europeum, Viola odorata*) was higher on both borders than on both territories, such a character of seedling distribution may support the hypothesis on nutrient-enriched habitats. Concentration of myrmecochore seedlings on territory borders is a result of centrifugal flow of diaspores directed from the nest to the territory borders (Appendix, Figs. XI; XII, A, C, E). Presumably, higher content of nutrients in the border soil may also contribute to better germination of diaspores and higher survival of seedling in the first stages of development. Thus, the presence of garbage dumps, full of organic remains, on territory borders provides better conditions for plants, especially in the nutrient-poor habitats of the deciduous forest.

In a few experimental squares on the territory borders, a very high density of myrmecochore seedlings (*Viola odorata* and *Asarum europaeum*) was detected (320 – 1280 seedlings/m^2). Such squares were mainly located at borders located not far away from colony nests. At these microsites, an increase of competition between seedlings may be expected. However, seasonal dynamics of the border shape that depends on age and size of both colonies may increase scattering of secondarily relocated diaspores and lead to a decrease of competition between seedlings.

2. PLANT SPECIES COMPOSITION OF THE SOIL SEED POOL AT VARIOUS MICROSITES

Soil seed samples from different microsites are analysed in Table 7.3. A sample from the old border contained 883 diaspores/dm^3, from the new border – 402, from *L. fuliginosus* territory – 340, from *L. fuliginosus* nest – 737, from *F. polyctena* territory – 234, from *F. polyctena* nest – 392. In the soil samples, diaspores of a total of 22 plant species were detected, 16 of them in the sample from the new border, 17 – from the old border, 19 – from *L. fuliginosus* territory, 18 – from *F. polyctena* territory. In samples from *F. polyctena* and *L. fuliginosus* nests, diaspores of 14 and 20 species, respectively, were found; 9 of them were myrmecochores.

Comparisons of the soil seed pool at different microsites revealed concentrations of diaspores of three plant species (*Asarum europaeum, Viola odorata, Pulmonaria obscura*) on both borders. Four species (*Ballota nigra, Corydalis bulbosa, Anthriscus sylvestris, Geum urbanum*) were found only on the old border whereas three species (*Scilla bifolia, C. cava, Fallopia convolvulus*) on the new border only. Diaspore concentrations of *B. nigra, Chelidonium majus, F. convolvulus,* and *Glechoma hirsuta* were found in samples from nests of both ant species. Diaspores of *Corydalis bulbosa* and *Geum urbanum* tended to concentrate in the nest of *L. fuliginosus*. Diaspores of *Taraxacum officinale, Galium aparine, Anthriscus sylvestris, Omphalodes scorpioides, P. obscura,* and *Asarum europaeum* were concentrated in nests of *F. polyctena*.

Table 7.3. Seed material from ant nests, territories, and borders between territories of two ant species: *Formica polyctena* (FP) and *Lasius fuliginosus* (LF). Values correspond to the amount of diaspores in soil samples of 1 dm^3. Abbreviations: †, trees; *, myrmecochores.

Plant species	FP territory	LF territory	Old border	New border	LF nest	FP nest
* *Asarum europaeum*	3	1	241	7	0	13
* *Viola odorata*	26	47	183	191	11	38
* *Pulmonaria obscura*	0	0	14	8	0	6
* *Ballota nigra*	8	15	48	5	25	26
* *Chelidonium majus*	0	1	6	5	4	14
* *Corydalis bulbosa*	18	62	39	12	365	29
* *C. cava*	1	18	13	34	5	3
* *Omphalodes scorpioides*	3	1	3	3	3	12
* *Scilla bifolia*	0	4	6	27	0	2
Anthriscus sylvestris	16	14	99	3	0	44
Fallopia convolvulus	3	4	0	21	19	27
Geum urbanum	12	7	22	1	2	33
Glechoma hirsuta	3	4	5	0	31	18
Taraxacum officinale	2	0	1	4	4	21
Stellaria holostea	1	6	3	0	2	5
Anemone ranunculoides	0	0	0	4	0	0
Galium aparine	7	16	12	4	1	30
†*Ulmus laevis*	118	84	147	38	72	26
†*Robinia pseudoacacia*	3	6	5	5	0	15
†*Sambucus nigra*	0	0	0	0	0	1
†*Tilia cordata*	0	0	2	0	0	0
†*Acer platanoides*	10	57	28	29	6	29

Results of the PCA analysis for soil samples show that the first principal component accounts for 61.5% of variance, the second one for 24.3% (Fig. 7.3). Soil samples from territories of both ant species and from the *F. polyctena* nest form a rather compact cluster to the right of the middle part of the graph. The second cluster, located in the upper left part of the graph, contains samples from both borders. Sample from *L. fuliginosus* nest is located aside from other samples in the left lower part of the graph.

The first PC contributes mainly to the division of groups «territories» and «borders», whereas the second one mainly contributes to the division of groups «nests» and «borders». Three plant species (*Viola odorata, Corydalis bulbosa, Ballota nigra*) have the highest positive load on the first PC. *Asarum europaeum* and *V. odorata* have the highest positive load and *C. bulbosa* the highest negative load on the second PC.

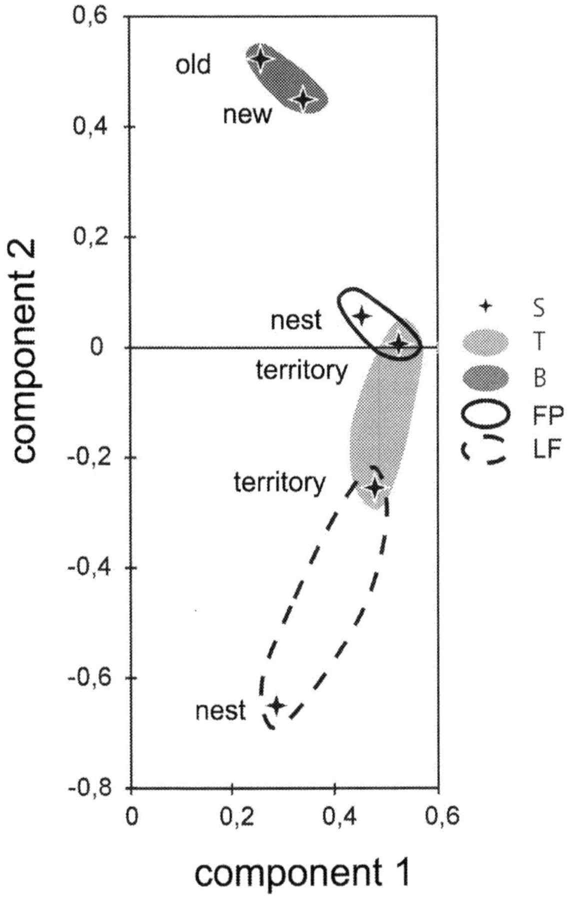

Fig. 7.3. Principal component analysis of soil samples. The first two component weights for seeds of nine myrmecochorous species from soil samples from six microsites. The first component accounts for 61.5% of variance, and the second one – 24.3%. B, border; FP, *Formica polyctena*; LF, *Lasius fuliginosus*; S, sample; T, territory.

3. DISTRIBUTION OF DIASPORES WITHIN TERRITORIES OF ANT COLONIES

3.1 Diaspore flow within territories of colonies of *Formica polyctena* and *Lasius fuliginosus* ants

Plant species abundance depends largely on the soil seed pool. Results of our analysis of the diaspore material from various microsites indicate the effects of species-specific features of ant behaviour as well as diverse plant diaspore dispersal strategies. Diaspore compositions of soil samples from nests and borders between territories were different from those of samples from either ant territory. Diaspores of some non-myrmecochores and small-seeded myrmecochores were concentrated in nests of both ant species, whereas diaspores of other non-myrmecochores and large-seeded myrmecochores were concentrated only nests of *F. polyctena*. The presence of a large amount of diaspores of non-myrmecochorous plants (in particular *Galium aparine* and *Anthriscus sylvestris*) in samples from nests may depend on the properties of the substrate of ant mounds, especially in *F. polyctena* ants. These plant species tend to concentrate on ant mounds and after fruiting leave an overwhelming majority of their diaspores in the mound substrate.

The amounts of diaspores of large-seeded (*Asarum europaeum, Viola odorata, Pulmonaria obscura, Scilla bifolia, Corydalis cava*) and some small-seeded myrmecochores (*Ballota nigra*) were somewhat higher in the samples from nests. Seeds of the small-seeded *Chelidonium majus* are concentrated in ant nests and very rarely transported by *F. polyctena* workers to territory borders. Even if they are transported, because of a high dropping rates of these seeds (Gorb and Gorb, 1999a), most of them do not reach the border and are dropped within the territory.

Seed samples from the nest and territory of *F. polyctena* ants are relatively similar. Samples from the nest and territory of *L. fuliginosus* show considerable difference. This may be explained by differences in foraging behaviour of each ant species: *F. polyctena* workers collect diaspores of all myrmecochores growing on their territory, whereas *L. fuliginosus* workers, presumably, prefer small diaspores (*Corydalis bulbosa, Ballota nigra, Chelidonium majus*). In the graph showing the results of the principal component analysis (Fig. 7.3), the sample from the *F. polyctena* nest has an intermediate position between territories of both ant species and borders of their colonies. This may be evidence for diaspore relocation from the nest to the border.

Diaspores of small-seeded myrmecochores are characterised by the higher dropping rates during transport by ants than the large-seeded myrmecochores. As previously mentioned (see chapter 5), such differences present a kind of filter, which increases the amount of dispersed diaspores in some plant species and decreases it in others during secondary relocation of diaspores.

3.2 Effect of ant behaviour on the distribution of diaspores

Previous authors that studied the phenomenon of myrmecochory focused mainly on clarification of various plant strategies (Beattie and Lyons, 1975; Handel, 1976; 1978; Beattie and Culver, 1981; Heithaus, 1986; Higashi et al., 1989; Ohkawara and Higashi, 1994). However, the foraging strategies of ant workers that collect seeds are also different (Traniello, 1989). Results presented in this chapter show that behavioural differences between ant species effect seed flow and seedling distribution in the deciduous forest. It has been previously reported that *Myrmica ruginodis* ants relocate diaspores of *Carex pilulifera* from their nests. This activity results in the secondary relocation of at least 16 – 20% of diaspores of this plant species (Kjellsson, 1985).

L. fuliginosus ants have a mainly carbohydrate diet and prefer aphid honey-dew and only use insects and elaiosome-bearing diaspores as an additional food source. From our observations, we know that ants of this species remove diaspores much more slowly and loose them more frequently than the carnivore *F. polyctena*. Garbage dumps of *L. fuliginosus* are located in the vicinity of nests, whereas *F. polyctena* workers transport wastes to the territory borders. This explains why it is exactly on the border between territories of these two ant species that such large amounts of both diaspores (Appendix, Fig. IX, D) and seedlings of myrmecochores were found.

Since the secondary diaspore flow within *F. polyctena* territories is directed to the territory border, diaspore-dropping during transport will provide additional diaspore scattering. Knowing that diaspore dropping rates during ant transport account for 8 – 10% of an initial seed amount on each meter of the route for large-seeded myrmecochores and for 20 – 100% for small-seeded ones, one might suggest that a considerable numbers of small diaspores do not reach the territory borders. Instead, they enrich the soil seed bank within the territory of this ant species. Because of a very low probability for small-seeded myrmecochores to reach a territory border, our data on these plant species do not support the nutrient-enrichment theory.

3.3 Advantages for plants from the secondary relocation of diaspores

On the basis of the data presented in this chapter, one may conclude that *F. polyctena* territories may provide additional advantages for large-seeded myrmecochores. The first advantage is the secondary relocation of diaspores by ants to territory borders. This leads not only to an increase of scattering rate of diaspores, but also promotes diaspore deposition in nutrient-enriched microsites. Furthermore, secondary relocation of diaspores prevents their circulation within the territory of the same colony, since on the borders, diaspores may be secondarily removed by ant workers from another ant colony or even other ant species. We have observed that diaspores deposited at the borders were often covered by a layer of organic waste material. This ant activity may lead to a better germination of diaspores at the borders. Previous authors reported that seeds germinate faster, if they are covered by a layer of ant waste. This phenomenon has been called «inhumation» (Beattie and Culver, 1982).

Diaspores of small-seeded myrmecochores often remain in ant nests, where the chances of germinating are quite low. On the other hand, they can be dropped within the territory of the colony without getting to the border. Therefore, these plant species have only the following advantages from the system with the secondary relocation of diaspores: dispersal for distance (Andersen, 1988) and escape from competition between seedling and parent plants as well as between seedlings themselves. In the next chapter, a computer simulation is applied to illustrate diaspore distribution for various dispersal systems by ants.

4. SUMMARY

- The chapter presents the results of field experiments on secondary relocation of diaspores from ant nests. Seed flow on ant territories, the floristic composition, distribution of mature plants and seedlings, and soil-seed material were studied. Distributions of plant material were analysed from ant territories and territory borders between colonies of the two ant species, *Lasius fuliginosus* and *Formica polyctena*. Additional seed samples were taken from ant nests.
- Workers of both species collect seeds of myrmecochores, but only individuals of *F. polyctena* relocate seeds to territory borders after the removal of elaiosomes. Borders of *F. polyctena* territories are thus

garbage dumps of waste organic material and probably also nutrient-enriched microsites.

- There were no great differences between the sampling sites in the abundance or the number of plant species (altogether 14 species) for mature myrmecochores or non-myrmecochores. Seeds of small-seeded myrmecochores seem to be more abundant in the nest of *L. fuliginosus*. Seeds and seedlings of myrmecochores were more abundant in territory borders than in other microsites.
- It is suggested that seed flow from nests of *F. polyctena* to territory borders results in an increase in the seed-dispersal distances from the parent plant in both small- and large-seeded myrmecochores. In large-seeded myrmecochores, this also results in a higher likelihood of reaching the «garbage dumps».

Chapter 8

Comparative analysis of plant dispersal systems by ants: diaspore concentration and redistribution

Previous chapters reported results of field experiments on diaspore dispersal by *Formica polyctena* ants. We were able to (1) determine removal rates of diaspores (Gorb and Gorb, 1995b; 2001a; Gorb, 1988), (2) estimate dropping rates of diaspores during their transport by ants (Gorb and Gorb, 1999a), and (3) evaluate directions of diaspore flows within colony territory (Gorb and Gorb, 1998b; Gorb et al., 2000). By using these experimental data, computer simulation may help to model dispersal process, in order to understand the effect of differences in the nest-cleaning behaviour of ants on spatial distribution of diaspores (seedlings).

The present chapter applies a computer simulation to analyse diaspore flows in an ant territory, spatial distribution of seedlings, and possible advantages of the secondary relocation of diaspores. A model is suggested for different species of myrmecochorous plants with various morphological types of diaspores and for two hypothetical ant species (one with a garbage dump located at nest entrances and another with waste deposited on territory borders).

1. VARIABLES USED FOR SIMULATION

The model environment contains the nest and colony territory (Fig. 8.1). The territory is circle-shaped with a radius TR. A parent plant is located within the territory and produces N diaspores.

Based on the data of the field experiments, co-ordinates of each ant-dispersed diaspore were determined as follows. 1) Ant workers transport diaspores to the nest within the constant sector S having an angle of 90°. 2) Some diaspores (L) are dropped (without secondary removal) by ants in each single distance of the route (DS). The single distance, within which L % of diaspores is dropped, is set at 3 m. This distance corresponds to that used in the field experiments on the evaluation of the dropping rates of seeds (Gorb and Gorb, 1999a). L clearly depends on morphological features of diaspores of particular plant species, and is therefore a species-specific variable. For small-seeded plants, L values are large and vice versa. 3) Diaspores, which are not dropped during transport, reach the nest.

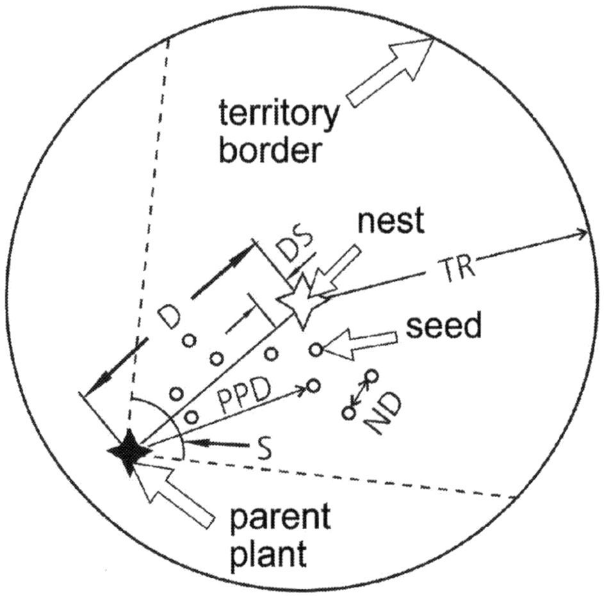

Fig. 8.1. Scheme of the set of factors. Black star, parent plant; D, distance from the parent plant to the nest; DS, single distance (3 m); large circle, territory border; ND, distance from one seed to the nest nearest seed; PPD, distance from a seed to the parent plant; S, sector facing the direction of the nest; small circles, seeds; TR, territory radius; white star, the nest.

Three variables were used for simulation: 1) the distance from the parent plant to the nest (D); 2) the percentage of diaspores dropped during transport within the single distance (L) and 3) the location of the waste deposited by ant workers. To evaluate the dependence of the dispersal distance of diaspores on the location of the parent plant relative to the nest, D was set from 1 to 16 m. D is constrained by the territory radius (16 m). To evaluate differences in the diaspore flow and spatial distribution of diaspores in plants with different morphological types of diaspores, L was set from 1% to 64%. To compare different systems of diaspore dispersal (with and without secondary relocation of diaspores), two variants were proposed.

Variant 1. After elaiosome removal, ants do not relocate diaspores to territory borders (*F. cinerea*-type). In this case, diaspores that reached the nest remain in the nest or in its vicinity. This represents a dispersal system without secondary relocation of diaspores.

Variant 2. Ants transport wastes to the territory border. This represents a dispersal system with secondary relocation of diaspores (*F. polyctena*-type). In this case diaspores that reached the nest are transported further by ants as follows. 1) The direction of the transport is randomly selected. 2) During secondary relocation of diaspores, ants drop diaspores more frequently (L=L+10%) than during the primary transport towards the nest. 3) In each single distance (DS=3 m), dropped diaspores are randomly distributed. 4) Diaspores, which were not dropped, reach the territory border.

According to hypotheses about selective advantages of myrmecochory, the most important parameters reflecting effectiveness of diaspore dispersal are an increase of the distance between seedlings and the greatest distance of seedlings from the parent plant (see chapter 9, section 1.1). Therefore, after simulation, distances to the nearest neighbour (ND) and to the parent plant (PPD) were determined for each simulated diaspore. Averages for ND and PPD were calculated on the basis of the simulation of 100 diaspores.

2. WHAT PROPORTION OF DIASPORES REACHES ANT NESTS?

The number of diaspores that reaches the nest primarily depends on the distance between the parent plant and the nest (Fig. 8.2, A) as well as on the dropping rate of seeds during transport (Fig. 8.2, B). The first variable is an individual characteristic of the parent plant, the second variable characterises plant species. In plants, located closer to the nest, a relatively large proportion of diaspores are deposited in the nest. In plants with a low dropping rate during transport by ants, there is also a considerable amount of diaspores reaching the nest. If the dropping rate is very low (L=1–4%), almost all diaspores are deposited in the nest. The highest difference between plants that grow at various distances from the nest, is revealed at L=16–32%. Dependence of the amount of diaspores reaching the nest from on dropping rate has an exponential character (Fig. 8.2, B).

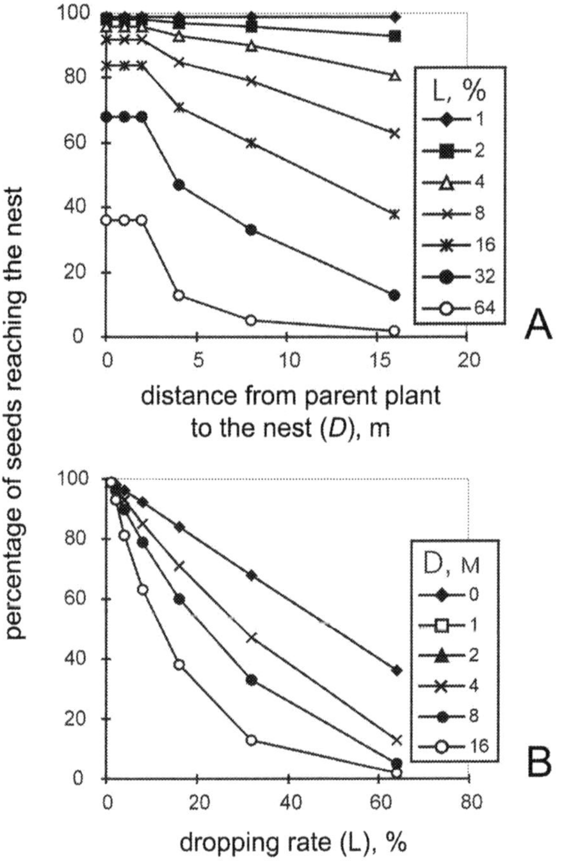

Fig. 8.2. Dependence of seed number that reached the nest on the distance of the parent plant from the nest (A) and on the dropping rate of diaspores (B).

3. COMPARISON OF SYSTEMS OF DIASPORE DISPERSAL BY ANTS

3.1 Dispersal without secondary relocation of diaspores

The main characteristic of this variant is the fact that all diaspores that reached the nest remain there. Furthermore, the distance between dispersed diaspores is very small. Comparison of cases for plants, located at various distances from the nest (D= 2 and 16 m) and those having various diaspore morphologies (L= 8% and 32%) revealed that almost all diaspores are deposited into the nest if parent plants grow in the vicinity of ant nests (Fig. 8.3, A, C). This results in escaping from the negative influence of the parent plant (Fig. 8.3, I, K), however, the distance between dispersed diaspores is very small (Fig. 8.3, E, G). There was not a considerable difference between plants with different values of dropping rates. Diaspores of plants located far from the nest are partly dropped on colony territory (Fig. 8.3, B, D). Therefore, only a small amount of diaspores reaches the nest. In this case, distances between dispersed diaspores and between the diaspores and the parent plant tend to increase (Fig. 8.3, F, H, J, L).

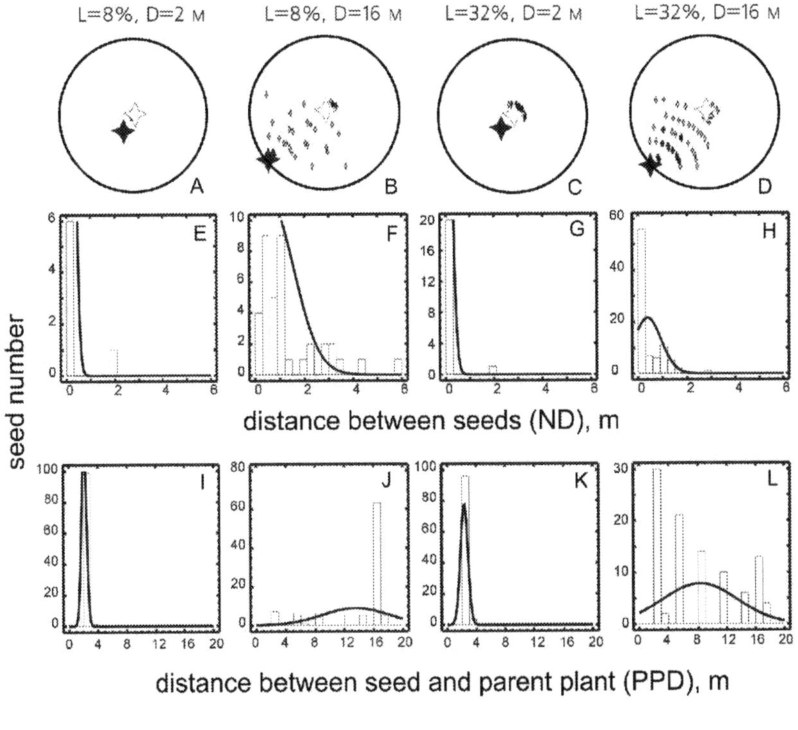

Fig. 8.3. Dispersal system without secondary relocation of diaspores. A – D. Computer simulated patterns of diaspore distribution of parent plants with different values of plant distance from the nest and dropping rates of diaspores. E – L. Frequency histograms showing the distance from the diaspore to the nearest diaspore (E – H) and from the diaspore to the parent plant (I – L) for examples presented in A – D.

Dependence of the distance between dispersed diaspores on the dropping rates differs in plants located at various distances from the nest (Fig. 8.4, A).

At a closer distance (up to 8 m), dropping rates have only a minor effect on the distance between dispersed diaspores. At a greater distance, there is a peak of distances between dispersed diaspores at L=8%. If the dropping percentage further increases, the distance between dispersed diaspores and parent plant does not change or slightly increases for parent plants located close to the nest and abruptly decreases for plants located far from the nest (Fig. 8.4, B).

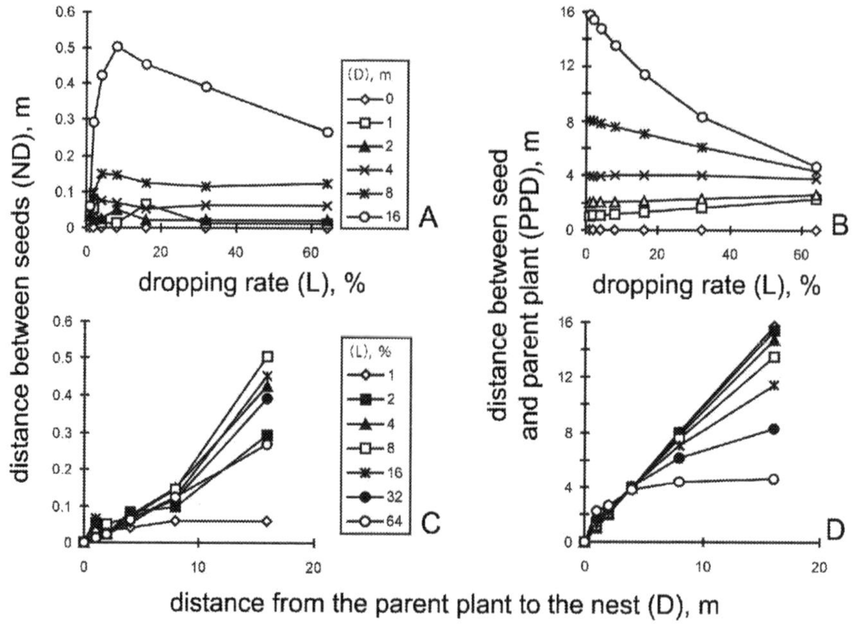

Fig. 8.4. Dispersal system without secondary relocation of diaspores. Dependence of the distance from the diaspore to the next nearest diaspore (A, C) and from the diaspore to the parent plant (B, D) on dropping rates (A, B) and on the parent plant distance from the nest (C, D). Each point is an average calculated for 100 seeds.

Dependence of the distance between dispersed diaspores on the distance of the parent plant from the nest is similar in plants having similar dropping rate values. A further increase of the distance between the parent plant and the nest usually leads to an increase of the distance between dispersed diaspores. However, at low dropping rate values, the distance between dispersed diaspores remains constant (Fig. 8.4, C). Dependence of the distance between diaspores and parent plant on the plant's distance from the nest has a logarithmic character at high dropping rate values and becomes almost linear at low ones (Fig. 8.4, D). If the distance of the parent plant to the nest increases, the distance between dispersed diaspores and the parent plants also increases.

Dispersal systems without secondary relocation of diaspores are most effective for a plant, when the plant is located at a relatively greater distance from the nest (D=4 – 16 m) (Fig. 8.5, A, B). Plants with dropping rates of 5 – 10% have better chances for diaspore dispersal than plants with very high or very low dropping rates.

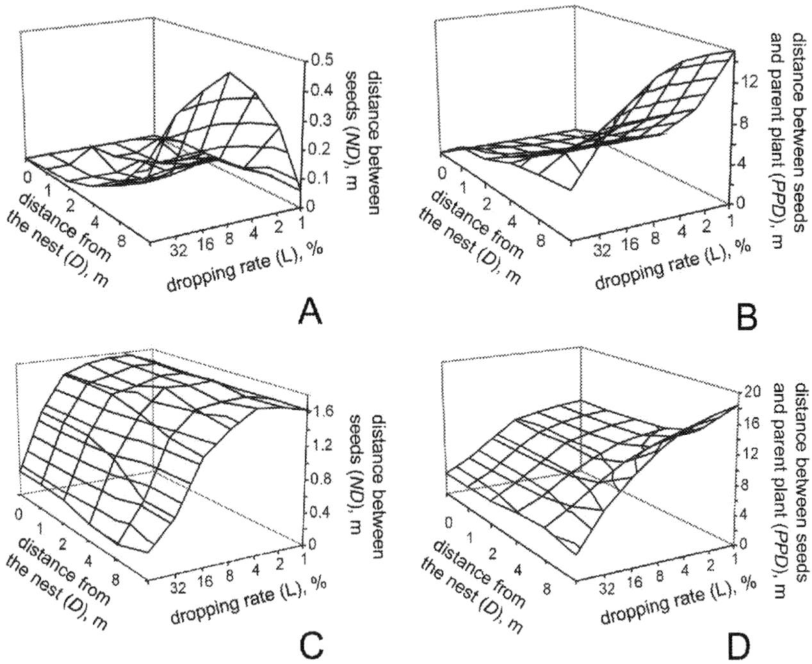

Fig. 8.5. Dependence of the distance between dispersed diaspores (A, C) between the diaspores and the parent plant (B, D) on the plant distance from the nest and on dropping rates in systems with secondary relocation of diaspores (C, D) and without it (A, B). Every point is an average calculated for 100 seeds.

3.2 Dispersal with secondary relocation of diaspores

In this case, diaspores that reached the nest will be transported by ants to the territory border. Thus, the entire dispersal process has two stages of dispersal. In the first stage, the diaspore flow is directed to the nest and constrained by the sector S. In the second stage, each diaspore is transported from the nest to the territory border in an arbitrary direction.

From the simulation of the dispersal system without secondary relocation of diaspores, it is known that almost all diaspores from plants, growing in the vicinity of the nest, reach the nest. However, in the case of the system with relocation, diaspores do not concentrate in the nest. This leads to the disappearance of the difference in the distance between dispersed diaspores in plants located at various distances from the nest (Fig. 8.6; compare E with F and G with H). Distance between diaspores tends to increase with the decreased values of dropping rates (Fig. 8.6; compare E with G and F with H). The resulting distance between diaspores and parent plant is greater, if the parent plant is located at a greater distance from the nest, and the dropping rate is low (Fig. 8.6, I – L).

Diaspores of parent plants that are located far from the nest are partly dropped in the ant territory during primary transport. In this case, a lower amount of diaspores are deposited in the nest, from where they would then be secondarily relocated by the ants.

For parent plants located at different distances from the nest, dependence of the distance between dispersed diaspores on dropping rates remains constant (Fig. 8.7, A). Distance between diaspores considerably decreases, if the dropping rate increases. Dependence of the distance between diaspores and parent plant on the dropping rates has an exponential character: the distance decreases with an increase of the dropping rate (Fig. 8.7, B). Distance between dispersed diaspores slightly decreases with an increase of the distance between the parent plant and the nest (Fig. 8.7, C). At a very low dropping rate, this relationship has an almost linear character. Remoteness of diaspores from the parent plant increases with an increase of the distance between the plant and the nest. However, at a very high dropping rate (>64%), the distance between the parent plant and dispersed diaspores decreases (Fig. 8.7, D).

Dispersal systems with a secondary relocation of diaspores are effective independently of the remoteness of the parent plant from the nest (Fig. 8.5, C, D). Plants with L<20% have better diaspore dispersal than plants with higher dropping rate values.

4. ANALYSIS OF DIASPORE DISPERSAL SYSTEMS

Previous authors that studied various aspects of myrmecochory did not consider the influence of species-specific features of ant behaviour on diaspore dispersal. There are only a few reports about the secondary relocation of diaspores. After elaiosome removal, *Myrmica ruginodis* ants transport diaspore bodies of *Carex pilulifera* from the nest (Kjellsson, 1985). *Formica rufa* s. l. workers deposit diaspore bodies on their territory border, whereas other species from the genus *Formica* do not do this (Dlussky, 1967). Computer simulation made possible the evaluation of spatial distribution of diaspores for the two types of dispersal systems by ants. Its results show that the system with the secondary relocation of diaspores to territory borders is more advantageous for myrmecochorous plants of the temperate deciduous forest than the system without it, because of higher dispersal distance and higher degree of diaspore scattering.

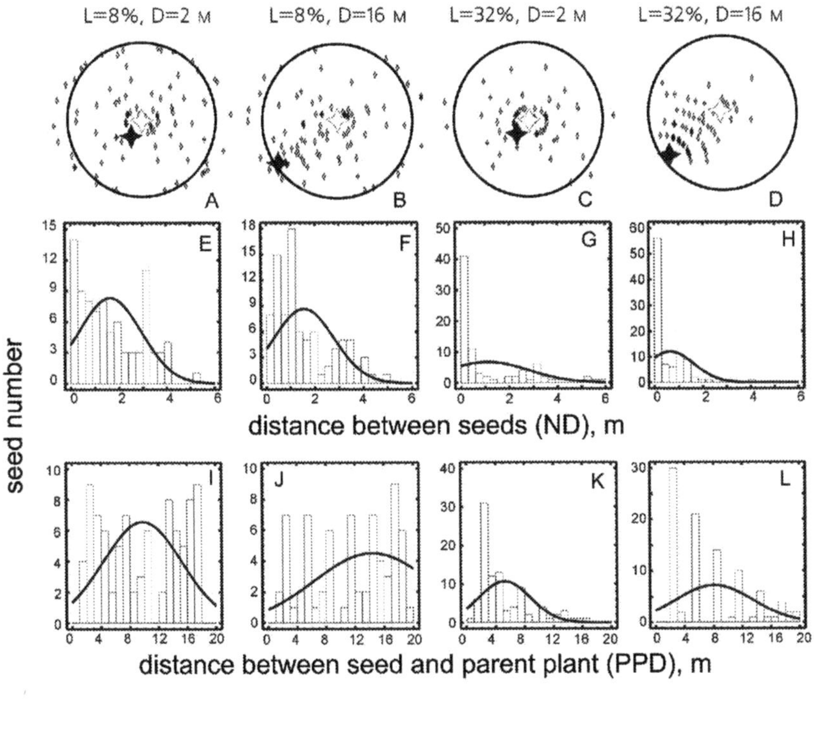

Fig. 8.6. Dispersal system with secondary relocation of diaspores. A – D. Computer simulated pattern of the diaspore distribution for plants with different distances from the nest and dropping rates of diaspores. E – L. Frequency histograms showing the nearest distance between the diaspores (E – H) and between each diaspore and the parent plant (I – L) for examples presented in A – D.

Diaspore concentration in nests is advantageous for plants to avoid fires and to escape diaspore predation by rodents and ground beetles (Beattie,

1983; Handel and Beattie, 1990). Since fires in the forest studied are a very rare event, diaspore deposition into the nest chambers located deeply in the soil does not provide any advantages for myrmecochores.

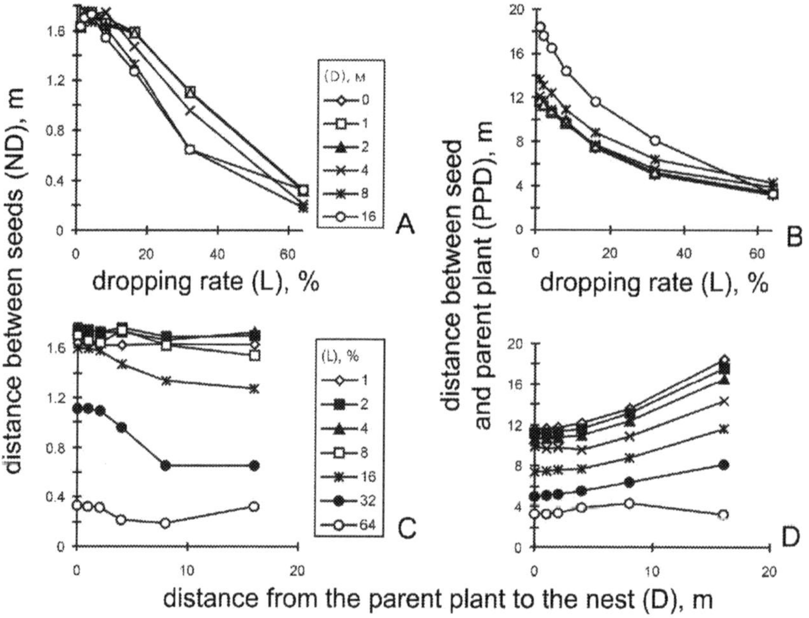

Fig. 8.7. Dispersal system with secondary relocation of diaspores. Dependence of the distance from one diaspore to the next nearest diaspore (A, C) and from the diaspore to the parent plant (B, D) on the diaspore dropping rate (A, B) and on the parent plant distance from the nest (C, D). Every point is an average calculated for 100 seeds.

It is well known that the nest substrate has a higher nutrition content than the surrounding soil and some physical and chemical features, which provide better conditions for diaspore germination and seedling growth (Greenfeld, 1941; Czervinski et al., 1971; Malozemova and Koruma, 1973; King, 1977a; b; c; Handel, 1978a; Petal, 1978; Culver and Beattie, 1980; Davidson and Morton, 1981a; Beattie, 1983; Beattie and Culver, 1983; Handel and Beattie, 1990) (Appendix, Figs. VIII; XIII, A, D). Therefore, many mature myrmecochores, such as *Corydalis* spp., *Scilla* spp., *Gagea* spp., *Viola odorata, Lamium maculatum, Chelidonium majus*, and their seedlings are often observed on the mounds of *F. polyctena* ants. On the other hand, ants may kill plants growing on their nests by covering them with soil or biting their roots (see chapter 10). Moreover, a high concentration of diaspores may lead to a high competition between seedlings. Diaspores of some myrmecochores, in particular those of *Asarum europaeum* and *Pulmonaria obscura*, do not germinate in nests of *F. polyctena*, although a large amount of them was found in the nest substrate. It may be concluded that some plant species obtain advantages from their deposition in the nest and others do not.

Computer simulations presented above show that in the case of the dispersal system without secondary relocation, diaspores tend to concentrate in ant nests. This might lead to the increase of the plant density on nests, and consequently to the increase of the competition between them. This dispersal system may provide some advantages for myrmecochores in ecosystems with frequent fires, high pressure of seed predators, and extremely low nutrient content of the forest soil. As to dispersal for distance, the system without secondary relocation provides quite reasonable diaspore scattering for plants located far from the nest. Many myrmecochores, such as *Ch. majus, V. matutina, Ballota ruderalis, L. maculatum*, produce a high number of small diaspores with high dropping rates during transport by ants. Obligate myrmecochores produce large diaspores with low dropping rates. Myrmecochores with middle-sized diaspores with intermediate dropping rates have better results of the dispersal in distance than small-seeded (large values of L) and large-seeded (small values of L) myrmecochores.

On the other hand, the systems with secondary relocation provide a two to three fold higher distance between dispersed diaspores and more regular distribution of diaspores in an area, independent of the plant distance from the nest. Large-seeded myrmecochores are dispersed more effectively than small-seeded ones. Experimental data on *A. europaeum* support this result of the simulation: distance between seedlings and parent plant was higher and seedling mortality was lower at the microsites in the deciduous forest with *F. polyctena* nests (see chapter 9, section 4). Moreover, the system with secondary relocation may prevent diaspore circulation within the territory of the same ant colony. Diaspores of plants that grew on territory borders from the diaspores previously deposited there by ants, have a higher probability of being dispersed by ant workers from the neighbouring ant colony.

Described mechanisms may influence plant species composition in various microsites of the deciduous forest. Although the model takes into account only a limited number of specific factors, it demonstrates relatively well the different systems and strategies of diaspore dispersal by ants. After obtaining new experimental data, it would be possible to enhance the model by including a higher number of factors and situations.

5. SUMMARY

- In this chapter, the seed flows within ant territories, the seed distribution, and benefits for myrmecochores with different morphological types of diaspores from secondary diaspore relocation are analysed.
- Using computer simulation, the effect of differences in ant behaviour during the relocation of waste materials on the spatial pattern of diaspores dispersed is shown. By varying the parameters describing parent plant location and dropping rate of diaspores, it is determined how variations in two main types of dispersal systems result in a spatial pattern of diaspores transported by ants.
- The results indicate that the myrmecochores may have an additional benefit in territories of ant species which relocate waste material to the territory borders. A dispersal system with secondary diaspore relocation provides a greater distance between diaspores and the parent plant as well as between diaspores, compared with systems without diaspore relocation.

Chapter 9

Ecological implications of myrmecochory

Two fundamental criteria of myrmecophily (mutual interactions between ants and plants) are generally implicated (Blatter, 1928). First, plants have to obtain evident advantages from the ant activity, and ants have to be able to provide these advantages. Second, plants must have specialised structures for ant attraction. These criteria are also fully applicable to myrmecochory. Since the appearance of the first monograph on myrmecochory (Sernander, 1906), a lot of information has accumulated on the second criterion (anatomy and biochemistry of elaiosomes). As to the first criterion, five hypotheses of selective advantages of plant dispersal by ants have been proposed (see chapter 1, section 9). This chapter briefly presents the main claims of the hypotheses and provides an analysis based on the data from the literature and a few case studies targeted to prove some of the hypotheses.

1. SELECTIVE ADVANTAGES OF MYRMECOCHORY

1.1 Hypotheses of selective advantages of myrmecochory

1.1.1 Hypothesis of the nest environment

Experiments with two species of the genus *Viola* have demonstrated that after seed removal, transport, and deposition into the nest, seeds germinate within the nest and seedlings appear on the mound (Culver and Beattie, 1978). Later it was found that seedlings are larger, their density is higher, and mortality considerably lower on the mounds than in surrounding habitats (Culver and Beattie, 1980). These results may be explained by a higher content of organic substances and higher nutrition content of the nest soil. Nest substrates of representatives from the genera *Myrmica, Lasius, Formica* and other active seed collectors contain a considerably higher concentration of nutrients, nitrogen, potassium, phosphorus, as well as alkaline cations than that of the surrounding soil (Czervinski et al., 1971; Malozemova and Koruma, 1973; Horvitz and Schemske, 1986a). Nest soil is looser, better aerated and is able to retain humidity (Handel and Beattie, 1990). Since ants provide a stable environment for their colony members (Petal, 1978), seedlings on the nests are also protected from the extreme impact of environmental factors.

This selective advantage is especially important in microsites with nutrient limited soil (Jansen, 1974; Westoby et al., 1990a), such as forest ecosystems, where herbaceous vegetation lacks a sufficient amount of nutrition due to the accumulation of nutrients in tree biomass (Grime, 1977; Muller, 1978; Beattie and Culver, 1981).

1.1.2 Hypothesis of diaspore escape from predators

Diaspores of myrmecochores may be intensively eaten by small rodents, birds, ground beetles, and other animals (Turnbull and Culver, 1983; Higashi et al., 1989; Ohkawara et al., 1997). Experiments with forest and desert myrmecochores have shown that if diaspores are not removed from fruits fast enough by ants, they are normally destroyed by seed predators (O'Dowd and Hay, 1980; Heithaus, 1981). Removal and deposition of diaspores into ant nests save them from destruction by some granivorous animals. Adherents of this hypothesis believe that diaspore eating by granivores was the main driving force contributing to the natural selection of plant species towards myrmecochory.

1.1.3 Competition avoidance hypothesis

Experiments with three species from the genus *Carex* have shown that myrmecochory leads to a decrease of the competition between closely-related species of plants (Handel, 1976; 1978a). One of species studied is a myrmecochore, whose seeds escaped competition with other species when they were deposited by ants into their nests. The experiments provided evidence that the growth and flowering of individuals of this species were suppressed, if they grew in the vicinity of non-myrmecochores.

Earlier, many other pairs of closely-related species of myrmecochore-non-myrmecochore have been identified (e.g. see Fig. 1.8, chapter 1)

(Ulbrich, 1928; Uphof, 1942). These numerous examples support the general opinion that this selective advantage is wide spread.

1.1.4 Fire escape hypothesis

In the Californian chaparral and Australian sclerophile forests, where periodical fires are a substantial ecological factor, diaspores of myrmecochorous plants, deposited in ant nests, are protected from overheating (Berg, 1975). The presence of a large number of myrmecochores in these habitats led previous authors to suggest that behaviour of ants transporting seeds plays an important role in diaspore dispersal and survival in these habitats. Moreover, seeds of many myrmecochorous Fabaceae from western Australia even require fire stimulation for successful germination (Shea et al., 1979). In the last case, seeds have to be hidden in the soil, not very deeply (to be warmed enough), and at the same time not too close to the surface (to avoid lethal overheating) (Beattie, 1983). This selective advantage of myrmecochory plays the most important role in Australian and South African shrub ecosystems (Handel and Beattie, 1990).

1.1.5 Hypothesis of the dispersal for distance

Studies of Australian myrmecochores showed an interesting rule: if the seedling mortality at the early developmental stages is inversely proportional to the distance at which seeds were transported from the parent plant, ants might increase the number of survived seedlings by means of seed transport to some optimal distance from the parent plant (Westoby and Rice, 1981). Initially, diaspores may also be dispersed by other mechanisms, such as autochory, but seedling viability depends on the further relocation of diaspores far from the parent plant and/or from the population in general.

The main advantage for myrmecochorous plants from dispersal by ants is an expansion of the borders of their natural habitats. Usually, ants transport diaspores for only several meters, however, a distance up to 70 m were registered (Sernander, 1906). Thus, plants obtain a chance for occupation of new habitats. Generally, population scattering decreases the extinction probability due to local changes of environmental factors. This advantage may be provided by almost any ant species independent of its behavioural features (Handel and Beattie, 1990).

1.2 Recent studies of selective advantages of myrmecochory

Many studies on myrmecochory that were published in the last 15 years provide evidence supporting or rejecting the above hypotheses. For example, the hypothesis that plants reach nutrient-enriched habitats due to diaspore dispersal by ants has been supported for several plant species (Oostermeijer, 1989; Danin and Yom-Tov, 1990; Dean and Yeaton, 1992; Gibson, 1993a; b; Griffith, 1996; Gorb et al., 1997; Sindarovskaya et al., 1997). However, other authors have demonstrated that plants do not obtain any additional advantages from diaspore deposition into ant nests, in spite of the higher nutrition content of the nest soil (Horvitz and Schemske, 1986a; Rice and Westoby, 1986; Bond and Stock, 1989; Hughes, 1991).

Some authors collected much evidence supporting the hypothesis of predator avoidance (Turnbull and Culver, 1983; Heithaus, 1981; Ohara and Higashi, 1987; Higashi and Ito, 1991; Okhawara et al., 1996; 1997; Espadaler and Gomes, 1996). Other studies, however, showed that granivorous predators do not influence diaspore dispersal in *Calathea ovandensis* (Horvitz and Schemske, 1986b). For representatives of the genus *Melampyrum*, the main selective advantage of myrmecochory is not the escape from predators but rather an occupation of nutrient-enriched habitats (Gibson, 1993b). Some authors (Westoby et al., 1991a; b) obtained results which did not support either hypotheses, the nutrient-enriched habitats or the predator avoidance for African and Australian plants growing on an extremely poor soil.

The hypothesis of the avoidance of competition between seedlings has been clearly supported in the myrmecochore *Trillium tschonoskii* (Higashi et al., 1989). Australian plants *Acacia terminalis* and *Dillwynia retorta* have an advantage from fire avoidance through myrmecochory (Hughes and Westoby, 1992b). Generally, for region of Australia, the most widely spread advantage of myrmecochory is dispersal for distance (Andersen, 1988).

2. ANT NESTS AS MICROHABITATS FOR MYRMECOCHORES

The nest environment hypothesis (Handel, 1978a; Culver and Beattie, 1980; Beattie, 1983; Beattie and Culver, 1983) considers occupation of the nutrient-enriched habitats on ant nests as the main selective advantage of myrmecochory. However, other authors reject the universality of the hypothesis, arguing that only a minor portion of diaspores reaches the nutrient-enriched habitats (Kjellsson, 1985; Horvitz and Schemske, 1986a; Rice and Westoby, 1986; Bond and Stock, 1989; Bond et al., 1991; Hughes, 1991). Since the majority of seeds of many ant-dispersing plants have the advantage of seed relocation for some distance from the parent plant, many authors assume that the main selective advantage of myrmecochory is a decrease of the competition between seedlings and parent plant and between seedlings themselves (Andersen, 1988).

Our data on the soil seed pool of the nest substrate of *Formica polyctena* ants clearly demonstrated that the main part of diaspores of myrmecochores that reached the nest is secondarily relocated by ants and redistributed at microsites with poorer soil than the nest substrate (see chapter 6). The secondary relocation of diaspores is a result of a specific behaviour of ant workers transporting wastes of the colony to territory borders (Dlussky, 1967). In spite of such a cleaning behaviour, some portion of the diaspores remain in the nest and germinate there. What is the fate of these plants? What are the advantages of growth on ant nests?

Below, an experimental analysis of advantages for the plant *Viola odorata* growing on the mounds of *F. polyctena* nests is presented. Fruit dimensions and seed producing capacity were compared in plants from mounds and from the forest microsites located outside of the mound (for a detailed method description see chapter 11, section 4.2.8). Symmetry and degree of fruit submersion were taken into account. In a separate experiment, an ability to damage fruits of *V. odorata* was tested among various ant species.

2.1 *Viola odorata* plants on nests of *Formica polyctena* ants

V. odorata usually occupies boundary microsites of the forest. During the fruiting period (June – September), diaspores originate from cleistogamous and chasmogamous flowers. Fruit and seed morphology of this myrmecochore is completely adapted to this method of diaspore dispersal. Fruits are spherical or elliptical capsules (with 10 – 30 seeds) lying on the surface of the forest litter (Fig. 9.1), under it, or slightly buried in the soil. Seed producing capacity is approximately 50 – 100 seeds per plant.

In plants growing on ant nests, the number of capsules was 1.68 times smaller than in control plants from the forest. 200 plants from the control sites contained a total 191 capsules (0.955 capsules per plant), whereas 176 plants from ant mounds – only 100 capsules (0.568 capsules per plant) (Fig. 9.2, A). Amount of capsules submerged in the soil was considerably higher in plants from ant mounds (84% of the total amount of capsules collected from the mounds) than in control plants from the forest (3% of the total amount of capsules collected out of the mounds) (Fig. 9.2, B).

Fig. 9.1. Fruits of *Viola odorata*. A. Capsule (BL) and flower bud (FB) at the level of the forest litter. B. Two workers of *Formica polyctena* ants probing the capsule. SD, seeds.

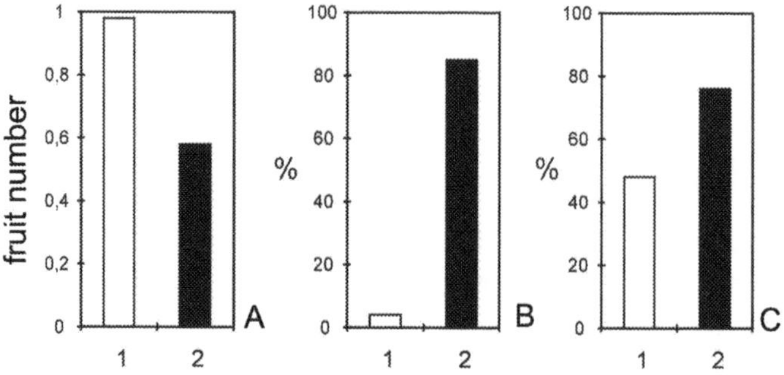

Fig. 9.2. Average number of fruits per plant (A), percentage of fruits submerged into the soil (B) and percentage of asymmetrical fruits (C) in *Viola odorata* plants from nest mounds of the *Formica polyctena* ant (black columns) and from control sites (white columns).

There were no considerable differences of in fruit size between plants from various microsites (length: on mounds – 8.56 mm, SD=1.31, n=41, forest – 8.42 mm, SD=1.07, n=37; width: on mounds – 7.49 mm, SD=1.38, n=41, forest – 7.39 mm, SD=0.96, n=37). However, the amount of asymmetric capsules was much higher in plants from mounds (71% of the total amount of capsules from the mounds) than in control plants from the forest (48% of the total amount of capsules from the control sites of the forest) (Fig. 9.2, C). The average amount of seeds in the capsules was not significantly higher in the control plants (x_{ave}=19.2, SD=6.32, n=37) compared with plants from the mounds (x_{ave}=17.6, SD=8.83, n=41).

Experiments, in which *V. odorata* capsules were exposed to various species of ants in the deciduous forest, showed that workers of *F. polyctena, Lasius fuliginosus*, and *L. niger* are able to damage the fruits. The highest portion of damaged fruits was found on the *F. polyctena* nests (Table 9.1).

Table 9.1. Results of experiments on the damage of *Viola odorata* capsules by various species of ants. bbreviation: n, number of recordings.

Ant species	Total number of capsules	Damaged	
	n	n	%
Formica polyctena	45	22	49
Lasius fuliginosus	40	9	23
Lasius niger	35	2	6
Lasius flavus	25	0	0
Myrmica rubra	45	0	0

2.2 Advantages for plants growing on ant nests

It has been previously reported that stemless violets from the section Nominium are highly adapted to dispersal by ants (Beattie and Lyons, 1975), and the removal of their seeds from the forest litter by *F. polyctena* workers is one of the highest among myrmecochores of the deciduous forest (Gorb and Gorb, 1995b). Therefore, these plant species are the most probable candidates to occupy microsites on the ant mounds. Comparison of various features of *V. odorata* fruits from the ants mounds and control microsites of the forest show that total seed producing capacity on the mounds is 1.5 times lower than elsewhere in the forest, in spite of the nutrient-richer habitats of ant mounds. The seed producing capacity in the forest was higher due mainly to a higher number of fruits per plant. Taking into account that the number of buried capsules was higher in plants from the mounds than in the control ones, one may suggest that a proportion of the capsules are covered with soil resulting from the nest-building activity of ants. Probably, covered capsules have a higher chance of escaping damage by ants: they have a similar amount of seeds as capsules of control plants. Thus, ability of *F.*

polyctena workers to damage *V. odorata* fruits leads to a decrease of the total seed producing capacity of plants on the mounds. A higher portion of asymmetric fruits in plants from the mounds is evidence that fruits are damaged in early stages of development.

Since ant activity is rather low in Spring, it may be suggested that such effects are lower in the early-spring ephemeral myrmecochores. Previous authors have demonstrated that in *Corydalis aurea,* a total amount of seeds in plants growing on ant nests was two times higher than in control plants (Handel and Beattie, 1990). Presumably, ephemeral myrmecochores, due to their rapid growth and development, may escape damage by ants that are inactive at that time. Additionally, fruits located in the upper part of the plant (*C. aurea*) are less subjected to damage than fruits lying on the soil (*V. odorata*).

Thus, microsites on the *F. polyctena* mounds may be advantageous to myrmecochores due mainly to physical and chemical properties of the soil. For some plants, however, such microsites may lead to a decrease in seed producing capacity due to the specific behaviour of the ant workers. Our data do not support the hypothesis of occupation of nutrient-enriched habitats as a main selective advantage of myrmecochory (Handel, 1978a; Culver and Beattie, 1980; Beattie, 1983; Beattie and Culver, 1983; Handel and Beattie, 1990) at least for mid-summer plants of the temperate deciduous forest and *F. polyctena* ants. Another example rejecting the hypothesis is the obligate myrmecochore *Asarum europaeum* which has never been found on *F. polyctena* mounds, in spite of a high abundance at the study site (Gorb and Gorb, 1995b) (see this chapter, section 4).

3. DECREASE OF SEEDLING DENSITY AS A SELECTIVE ADVANTAGE OF MYRMECOCHORY

Decrease of the competition between the parent plant and seedlings and between seedlings themselves is also one of the selective advantages of myrmecochory (Beattie, 1983; Handel and Beattie, 1990). Seedling mortality in the obligate myrmecochore *A. europaeum* is higher within the parent clone than outside of it (see this chapter, section 4). The advantage of seeds being relocated far from the parent plant was also experimentally supported for species from the genus *Trillium* (Mesler and Lu, 1983; Higashi et al., 1989).

Stems and pedicles of some myrmecochores droop or even lie down prior to dissemination (Ulbrich, 1928; Kjellson, 1985). In such a case, fruits are usually located just above the soil and empty their diaspores under gravity at a short distance from the parent plant (*autobarochory*) (for terminology see Levina, 1957). If this is a final stage of the dispersal process, seedlings will be located at a short distance from their parent plant as well as being close to each other and in high numbers. In the case of plants with elaiosome-bearing diaspores, the disseminated diaspores may be picked up by ants and dispersed further. In plants with the autobarochory as the first dispersal stage, the resulting spatial distribution of diaspores may differ from the distribution in obligate myrmecochores, whose fruits open at soil level.

One may suggest that even in the case of a very low density of ant workers at a microsite, diaspores of diplochorous (autobarochory + myrmecochory) species will be relocated at some distance from the parent plant. If ants are absent, passively disseminated seeds may be doomed to germination in groups. This will result in a high competition between seedlings. To verify this suggestion, spatial distribution of seedlings of the early-spring ephemeral myrmecochore *Corydalis bulbosa* (for methods see chapter 11, section 4.2.8) was analysed in a two-year experiment. This plant possesses a two-step dispersal mechanism, and has one of the earliest fruiting periods among ephemeral plants of the temperate deciduous forests (Gorb and Gorb, 1995a; Gorb, 1998). The experimental area (3.8×2.6 m) was divided into squares (approximately 10×10 cm), where previously, 16 mature plants of this plant species had been planted (Appendix, Fig. IX, E). In the following two years, orthogonal co-ordinates of each seedling were determined. With the aid of a designed computer program, the following distances were calculated. 1) The distance between each seedling to the closest parent plant. 2) The distance between each seedling to the closest seedling. 3) Distance from each experimental square with (or without) seedlings to the closest parent plant. *Lasius niger* ants were the only potential dispersal agent at the study site.

3.1 Spatial distribution of seedlings in *Corydalis bulbosa*

3.1.1 First year of the experiment

In the first year all seeds produced by plants were collected and counted to avoid their dispersal and germination in the following year. Therefore all variables, measured in the first year, reflect results of the spatial distribution of seeds in May of the previous year. An average seed producing capacity of plants at the experimental site was 82 seeds per plant (SD=18.4, n=16). A total of 355 seedlings germinated from seeds of the previous year were counted (Fig. 9.3, A). Taking into account the average seed producing capacity of plants, it amounted to about 26% of the expected number of seedlings. Presumably, the remaining 74% were seeds with a prolonged resting period (these fill up the soil seed bank), and loss due to disease and granivores.

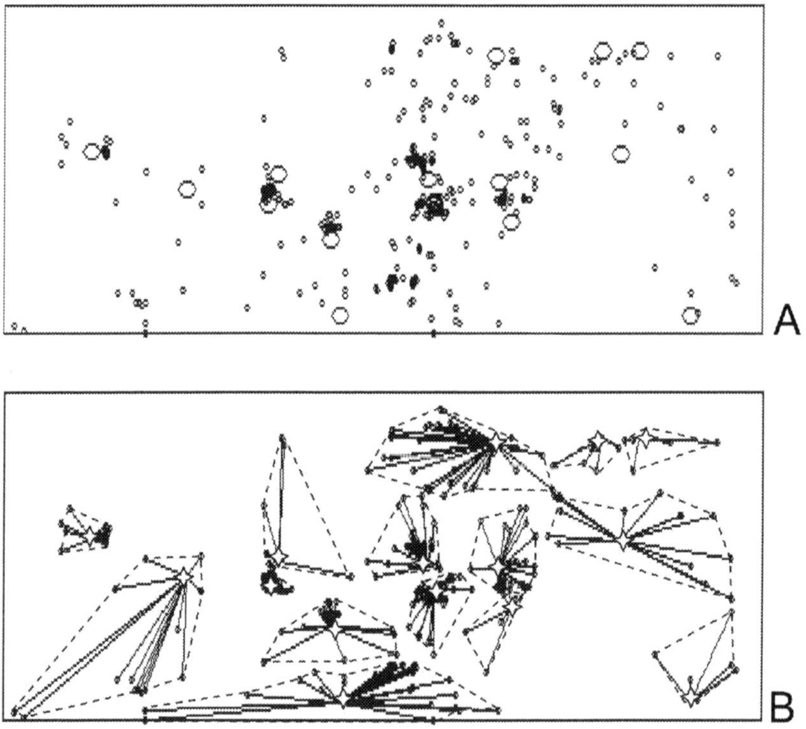

Fig. 9.3. *Corydalis bulbosa,* the first year of the experiment. A. Pattern of 355 seeds (small circles) and 16 adult plants (large circles). B. Distances between each seedling (small circles) and the nearest potential mature plant (stars).

The average distance between seedlings was 4.2 cm (SD=21.1, n=355). Out of 355 individually marked seedlings, 163 (46%) were located in groups, where the distance from each seedling to the next nearest seedling was practically zero (Figs. 9.3, A; 9.4, A). However, the majority of seedlings was located at a distance from mature plants (Figs. 9.3, B; 9.4, B). Only 30 seedlings (8%) were located at a distance smaller than 5 cm from the nearest potential parent plant. The majority of seedlings (n=141, 41%) were far away at 5 – 15 cm (x_{ave}=23.6 cm, SD=21.1, n=355) from the parent plant. Out of 988 experimental squares, 837 were without seedlings and 151 – with seedlings. Squares with seedlings were located significantly closer to the mature plants (x_{ave}=33.6 cm, SD=26.8, n=151) than squares without seedlings (x_{ave}=48.3 cm, SD=27.8, n=837) (Fig. 9.4, C, D).

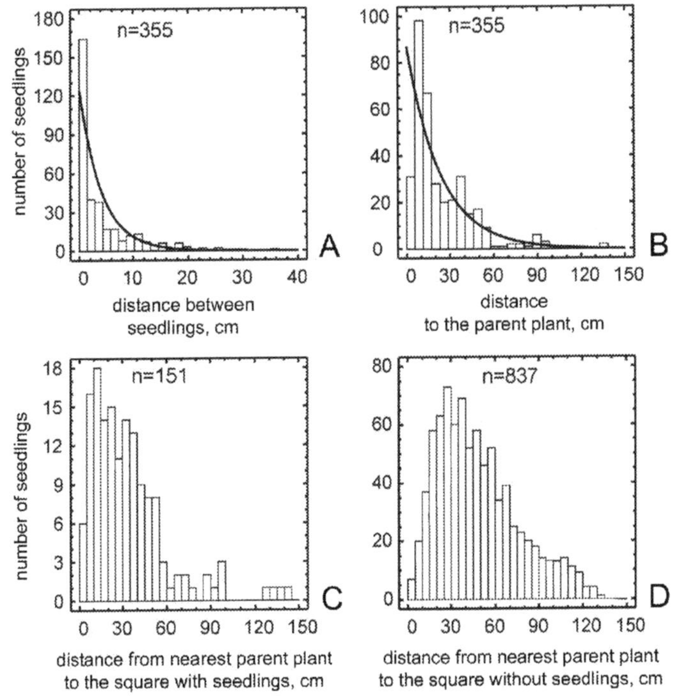

Fig. 9.4. *Corydalis bulbosa*, the first year of the experiment. Frequency histograms of the nearest distances between seedlings (A), between each seedling and the nearest potential mature plant (B), and between 988 squares (10×10 cm) and the nearest potential mature plant for squares with seedlings (C) and without them (D). Curves indicate exponential fits (A, B).

3.1.2 Second year of the experiment

In May of the second year of the experiment, 146 seedlings were registered at the experimental site (Fig. 9.5, A). Out of 355 seedlings, individually labelled the year before, 80 (23%) were found again. 66 seedlings presumably germinated from seeds resting in the soil seed bank.

The average distance between seedlings was significantly larger than in the year before (x_{cp}=7.3 cm, SD=8.9, n=146) (Figs. 9.5, B; 9.6, A). Average distance between seedlings and the nearest potential parent plant increased (x_{ave}=29.6 cm, SD=20.6, n=146 (Fig. 9.6, B). By subtracting the number of labelled seedling that survived (n=80) from the general sample (n=355), the number of dead seedlings was obtained (n=275). Comparison of the first and last samples showed that seedlings that survived were initially located significantly further away from other seedlings (x_{ave}=5.7 cm, SD=7.0, n=80) than dead ones (x_{ave}=3.7, SD=5.4, n=275). However, there was no significant difference in initial distance from each seedling to the nearest potential parent plant between these two groups of seedlings (survived: x_{ave}=24.7 cm, SD=20.8, n=80; dead: x_{ave}=23.2, SD=21.2, n=275).

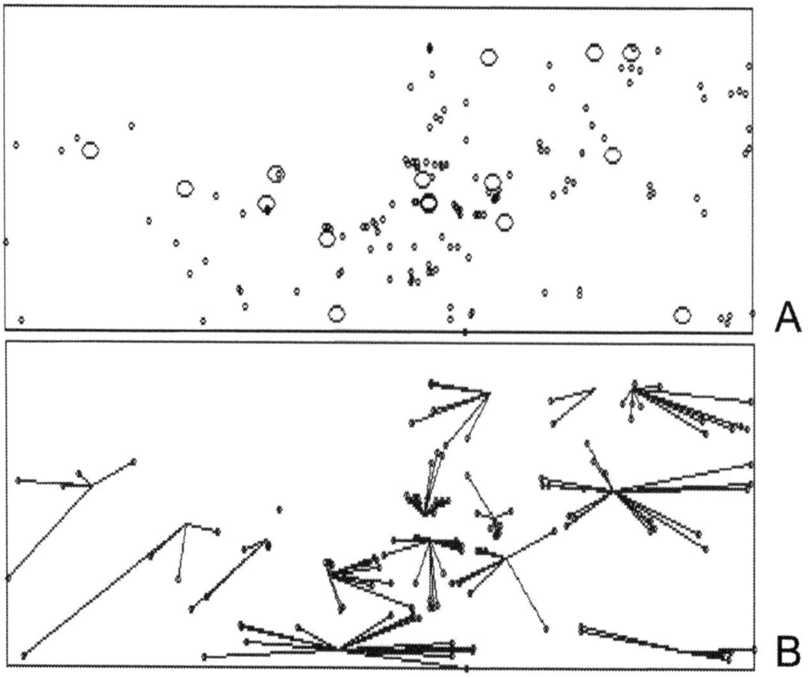

Fig. 9.5. *Corydalis bulbosa*, the second year of the experiment. A. Pattern of 146 seeds (small circles) and 16 mature plants (large circles). B. Distances from each seedling (small circles) to the nearest potential mature plant.

3.2 Effect of ants on seed dispersal of *Corydalis bulbosa*

In temperate deciduous forests, *F. polyctena* ants are the main agent of dispersal of elaiosome-bearing diaspores. Previous chapters (chapters 6 and

7) demonstrated that seeds of *C. bulbosa,* which are often observed as they are transported by ants towards their nests, do not concentrate in the *F. polyctena* nests. After removal and consumption of elaiosomes, seed bodies are transported by ants to territory borders. Seeds that remained in the nests germinated on the nest mounds. Due to a very rapid growth in the early Spring, when ants are weakly active, seedlings and mature plants of this ephemeral species are less subjected to covering by the soil than the slower growing seedlings of non-ephemeral plants (Gorb and Gorb, 1996). Advantages to growing on ant mounds have been previously demonstrated for other species from the same genus (*C. aurea*): total amount of seeds produced by plants from ant nests was twice as high as than in control plants (Handel and Beattie, 1990).

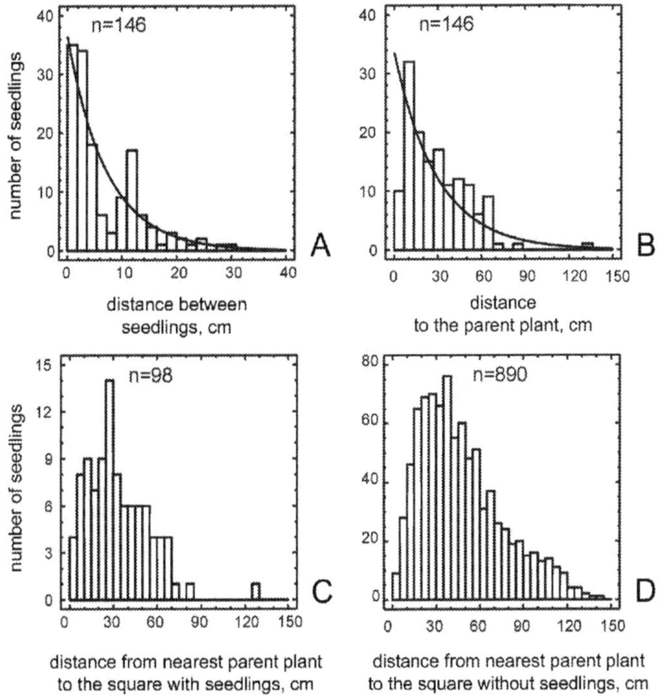

Fig. 9.6. *Corydalis bulbosa,* the second year of the experiment. Frequency histograms of the nearest distances between seedlings (A), between each seedling and the nearest potential parent plant (B), and between 988 squares (10×10 cm) and the nearest potential parent plant for squares with seedlings (C) and without them (D). Curves indicate exponential fits (A, B).

Our two-year experiment provided clear evidence that in *C. bulbosa*, a decrease of the pressure of the parent plant on seedlings is provided by the first stage of diaspore dispersal (autobarochory). Its result is the removal of seedlings to a distance of 10 – 15 cm from the parent plant. The presence of this dispersal stage ensures an important advantage for *C. bulbosa* seedlings, which escape from the negative effects of the parent plant. Seedlings of some other myrmecochores without the autochorous stage of dispersal die in the vicinity of parent plants at the microsites with a low density of ants (Gorb and Gorb, 1995c). Even seedlings far from the parent plant die in groups more often than solitary seedlings. The second (myrmecochorous) stage of dispersal is able to provide a decrease in competition between seedlings. Seeds of *C. bulbosa* have rather large elaiosomes and may be actively removed and transported by *F. polyctena* ants (Gorb and Gorb, 1995a; Gorb, 1998). Thus, each of these two dispersal stages provide particular advantages for a *C. bulbosa* seedling: the first one decreases the negative effects of the parent plant, the second one decreases the competition between seedlings.

Lasius niger ants, that were the only dispersal agents registered at the study site, are mainly specialised in collecting aphid honey-dew. Ants of this trophic specialisation are usually weakly attracted by the elaiosome content. However, they are able to provide carry diaspores away from groups, and increase seedling survival rate. The experimental situation described above corresponds to habitats at the marginal sites of the deciduous forest, where *C. bulbosa* and *L. niger* occurr together.

Diplochores, in particular *Chelidonium majus, Viola mirabilis,* and *V. matutina*, already solve both problems at the first stage of diaspore dispersal: decrease of the pressure of the parent plant and decrease of competition between seedlings. The myrmecochorous stage of dispersal is probably targeted in these species to reach nutrient-enriched microsites (Handel, 1978a; Culver and Beattie, 1980; Beattie, 1983; Beattie and Culver, 1983; Handel and Beattie, 1990). In diplochores, an insignificant role of myrmecochory for diaspore dispersal is reflected in the relationship between the elaiosome mass and seed mass(Gorb and Gorb, 1995b).

4. EFFECT OF MYRMECOCHORY ON THE SPATIAL DISTRIBUTION AND MORTALITY RATE OF SEEDLINGS IN *ASARUM EUROPAEUM*

One of the selective advantages of myrmecochory for plants is the diaspore dispersal for some distance from the parent plant (Beattie, 1983;

Handel and Beattie, 1990). As shown above, such a distance provides an occupation of new microsites by offspring as well as better conditions for growth and development of seedlings. For example, in *Trillium ovatum*, the negative effects of a parent plant decrease along with the decrease of the competition between seedlings (Messler and Lu, 1983). Sometimes even a short dispersal distance is sufficient to considerably reduce seedling mortality (Higashi et al., 1989).

For clone-forming plants, the distance of seedlings from the parent clone is a particularly important factor influencing seedling survival. In this case, the dispersal distance has to be enough to escape from the clone shade. The effect of ants on seedling survival in clone-forming plants was studied in the clone-forming obligate myrmecochore *Asarum europaeum* (Fig. 9.7). It was previously demonstrated (see chapter 2) that seeds of this plant have the highest removal rates by *F. polyctena* ants compared to other myrmecochorous plants of the forest.

Fig. 9.7. Typical clone of *Asarum europaeum* (A) and seedlings germinated from non-dispersed diaspores in the clone shade (B) .

There are microhabitats in which ant species composition and ant density vary. Which microhabitat is more advantageous for seed dispersal and seedling survival of *A. europaeum*? To answer this question, field experiments in two areas of the forest were carried out: with *F. polyctena* nests and without them (for methods see chapter 11, section 4.2.8). This study had two principal goals. The first was to compare spatial distribution of *A. europaeum* seedlings relative to the parent clone in forest areas with different density of *F. polyctena* colonies. The second was to estimate the mortality of seedlings located within the clone and out of the clone.

4.1 Spatial distribution and seedling mortality

In microsites without the presence of *F. polyctena* ants, the average number of seedlings in each circle of the clone was 5 – 10 times higher than in the microsite with nests of this ant species. This rule was found to apply as well for seedlings and for juvenile plants with true leaves (Fig. 9.8).

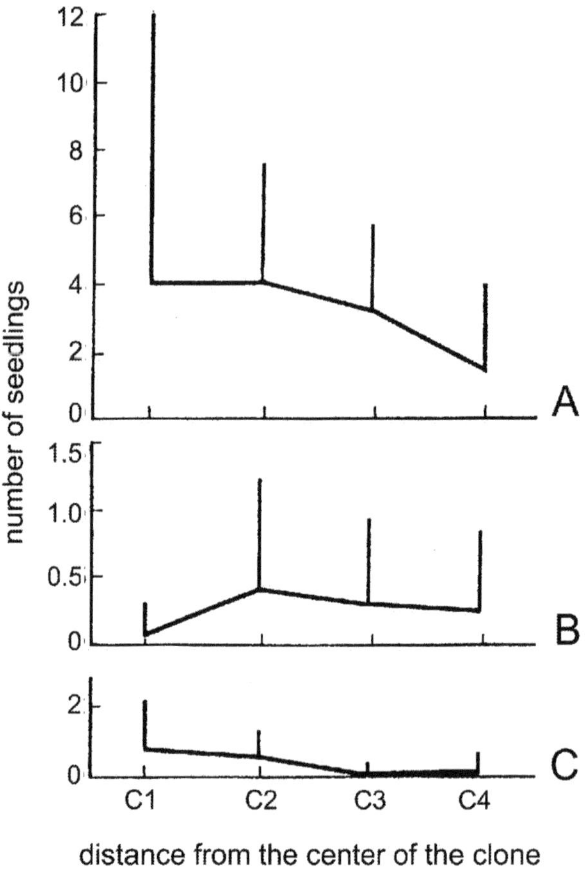

Fig. 9.8. Number of seedlings and juvenile plants of *Asarum europaeum* with cotyledons (A, C) and leaves (B). Plants are located at different distances from parental clones in areas with *Formica polyctena* nests (C) and without them (A, B). C1 – C4, four levels of seedling locations (see Fig. 11.7, chapter 11).

At the microsite with the ant nests, almost no juveniles were found in the vicinity of clones. However, samples obtained at the microsites studied have high standard deviations, which tend to decrease in the series of samples C1 – C4 (Fig. 9.8). In spite of such high deviations, numbers of seedlings are significantly different at various distances from the clone centre.

In the randomly selected squares (1 m^2) located outside of the clones, the average density of seedlings was 0.70 m^{-2} at microsites with nests (n=33) and 0.96 m^{-2} at microsites without nests (n=27). Thus, seedling density at microsites with nests of *F. polyctena* ants was lower than at microsites without nests.

Ratios between seedlings that survived or died during Winter were considerably different in clones than outside of them. In the sample of seedlings from outside the clones, the number of seedlings that survived was four times higher than the number of dead ones. In the sample of seedlings from the clones, the number of dead seedlings is four times higher than the number of those that survived (Fig. 9.9).

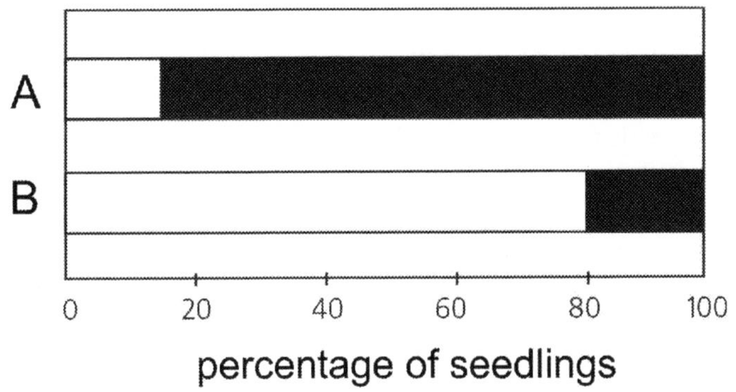

Fig. 9.9. Seedling mortality in *Asarum europaeum* inside clones (A) and outside of them (B), n=100. White bars, survived; black bars, dead.

4.2 Effect of ants on seed dispersal in *A. europaeum*

Shady microsites inside the clone of *A. europaeum* probably lead to the reduction of the photosynthetic activity of seedlings. This is reflected in the light-green colour of the leaves of seedlings inside clones. Location of seedlings inside clones is lethal: such seedlings perish in Autumn prior to the beginning of the cold season. Thus, myrmecochory in *A. europaeum* is vital for survival. Since *A. europaeum* seeds are very attractive to ants, have high removal rates by *F. polyctena* ants and can be transported a considerable distance (Gorb and Gorb, 1995b; 1999a), the majority of seeds are presumably deposited into the nest. This should be especially well expressed with high densities of ant workers and at short distances of the parent plant from the nest. However, neither clones nor single plants of *A. europaeum* were found on ant mounds. There are two explanations for such a situation. 1) Presumably, seeds do not germinate in nests. 2) Seeds do not concentrate in nests and are transported to territory borders. This is probably the reason why the seedling density in the randomly chosen squares was lower at microsites with *F. polyctena* nests than at microsites without nests. Results of our field experiments revealed that even a minimal distance from the parent clone is enough to drastically decrease seedling mortality. These results are well in accordance with data for *T. tschonoskii* (Higashi et al., 1989).

Interestingly, the majority of seedlings were located outside of the clones also at microsites without nests. Probably, even a minor density of *F. polyctena* individuals may provide a minimal dispersal of seeds from clones. Seed dispersal may also be provided with the aid of other ant species, such as *Myrmica rubra* and *Lasius fuliginosus,* which are present in the forest. Individuals of *M. rubra* are not able to transport seeds over long distances, but can move seeds for a distance of several centimetres.

According to the hypothesis of dispersal for distance, seedling dispersal for some distance from the parent plant provides better chances for seedling survival. Thus, results of the spatial distribution and mortality of *A. europaeum* seedlings support the hypothesis.

5. SUMMARY

- According to one of the fundamental criteria for myrmecophily, the plant must derive distinct advantages from the activities of ants. Currently there are five hypotheses for selective advantages of myrmecochory: (1) the nutrient-rich nest environment hypothesis, (2) the predator-avoidance hypothesis, (3) the competition-avoidance hypothesis, (4) the fire-avoidance hypothesis, and (5) the dispersal for distance hypothesis.
- In this chapter, the above hypotheses are briefly presented and the significance of some hypotheses for particular plant species are discussed on the basis of field experiment results.
- The comparison of fruit size, number, asymmetry and submergence into the soil, and seed producing capacity of the plants *Viola odorata*, growing on and out of nest mounds of the *Formica polyctena* ant,

supports the nutrient-rich nest environment hypothesis. The fruit number is lower and the number of asymmetrical and submerged fruits is higher in the plants from nest mounds, while fruit size and seed number are not significantly different between plants from different microsites.

- The investigation of both the spatial distribution of seeds and its effect on seedling mortality in *Corydalis bulbosa* with a two-phase type of seed dispersal (autobarochory + myrmecochory) showed that this plant species uses autobarochory to reduce the pressure of a parent plant on seedlings. However, even if far away from the parent plant, high seedling density leads to an increase of seedling mortality. The myrmecochorous phase in seed dispersal of *C. bulbosa* reduces competition between seedlings, as has been stated previously in the competition-avoidance hypothesis.
- The study of spatial pattern and seedling mortality in *Asarum europaeum* in two areas of the deciduous forest, with nests of ants *F. polyctena* and without them, supports the suggestion that the transference of seeds by ants to some distance from a parent plant promotes better conditions for seedlings. The number of seedlings in clones and the density of seedlings outside the clones are higher in the area without nests compared with the area with nests. Seedling mortality within clones is significantly higher than outside. These results support the dispersal for distance hypothesis.

Chapter 10

Interactions between ants and non-myrmecochorous plants

Diaspore dispersal by carnivorous ant species has been studied mainly for plants specialised for myrmecochory due to the presence of elaiosomes on their diaspores (Beattie and Lyons, 1975; Culver and Beattie, 1980; Beattie, 1983; Kjellsson, 1985; Higashi et al., 1989; Handel and Beattie, 1990; Gorb and Gorb, 1995a; b; c; 1996; 1997; 1998a; b; 1999a; b; 2000; 2001a; b; Gorb et al., 1996; 2000; Gorb, 1988; 2000). In the case of such a specialised myrmecochory, true mutualistic interactions take place between plants and ants: ants use elaiosomes as an energetically valuable food source, and plants take advantage of dispersal of their diaspores by ants (Heithaus et al., 1980; Handel and Beattie, 1990; Cushman and Beattie, 1991).

It has been also reported that ants may remove and transport diaspores without elaiosomes (Sernander, 1906). In spite of very low removal rates, diaspore relocation took place even in such cases of non-specialised plants. However, the further fate of such diaspores has not been previously studied.

In the deciduous forest, many non-myrmecochores, such as *Anthriscus sylvestris, Fallopia convolvulus, Alliaria petiolata, Impatiens noli-tangere, Galium aparine*, tend to concentrate on the nest mounds of *Formica polyctena* ants (Appendix, Fig. VIII). One may hypothesise that such guilds are not accidental, but rather the result of particular interactions between ants and plants.

In the present chapter, results of an experimental study of such interactions are described. The non-myrmecochore *G. aparine* and *F. polyctena* ants were used as model objects. Removal rates of *G. aparine* diaspores were estimated in the «cafeteria» experiments. Amounts of diaspores in the soil samples from the mound substrate and from randomly selected microsites in the ant territory were counted. Distribution of seedlings and growth dynamics of plants on mounds and on surrounding sites were studied (for a detailed method description see chapter 11, sections 4.2.9 and 4.3). The main goal was to analyse the advantages obtained by plants and ants from such an interaction. The following questions were asked: 1) How do *G. aparine* diaspores reach ant nests? 2) Does the soil seed pool of the nest substrate differ from that of the forest soil? 3) Are there differences in the diaspore germination dynamics and growth on nest mounds and on surrounding sites?

1. INTERACTIONS BETWEEN THE NON-MYRMECOCHORE *GALIUM APARINE* AND *FORMICA POLYCTENA* ANTS

1.1 Removal and transport of diaspores

Galium aparine is an annual herbaceous species with a climbing stem. The fruiting is in July – October. The fruit disintegrates into two round-shaped mericarps with small hooks on its surface. Due to hook coverage, this species is considered to be a highly specialised epizoochore (Gorb and Gorb, 2002).

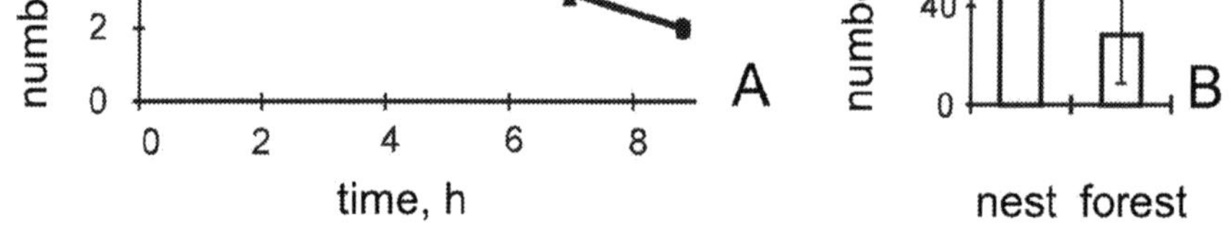

Fig. 10.1. Removal rates of *Galium aparine* diaspores by *Formica polyctena* ants (A) and diaspore number from soil sampled on the ant mounds and on surrounding sites (B).

«Cafeteria» experiments revealed that *F. polyctena* workers remove diaspores at a very low rate: approximately one diaspore per hour (Fig. 10.1,

A). Time of half removal is approximately five hours, which is much longer than for previously considered myrmecochores and diplochores (see chapter 2). Removed diaspores were usually dropped by ant workers at a distance of <20 – 30 cm from the seed depot. In spite of this, some diaspores reached the nest and concentrated there. Diaspore concentration is supported by the comparison of the soil seed pools: the substrate of the ant mound contains a higher amount of *G. aparine* diaspores than forest soil samples (Fig. 10.1, B).

1.2 Dynamics of seedling density and plant growth

Galium aparine diaspores start to germinate on the ant mounds in late August, and away from them in the late September. There is a considerable difference in the seedling number in the experimental squares on the mounds and in surrounding areas, particularly in the middle of the germination period (Fig. 10.2). Diaspores germinate in various parts of the mound (on the top, on slopes, on the base), but the highest density of seedlings is counted on mound slopes (30 – 60 cm from the mound top) (Fig. 10.3).

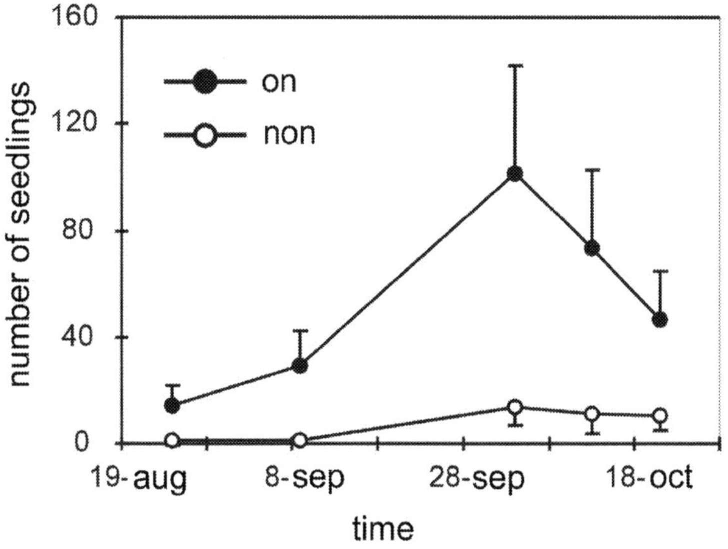

Fig. 10.2. Dynamics of seedling density in on-mound and non-mound experimental squares (25×25 cm) during the period of diaspore germination.

In early October, seedling density on the nest mounds start to abruptly decrease (Fig. 10.2). Moreover, the comparison of the seedling number in

October 1994 and April 1995 showed that almost all seedlings from the mound top (30 cm from the mound top centre) died during the Autumn and Winter (Fig. 10.3). Thus, the mound top and slopes are areas of the highest mortality of seedlings.

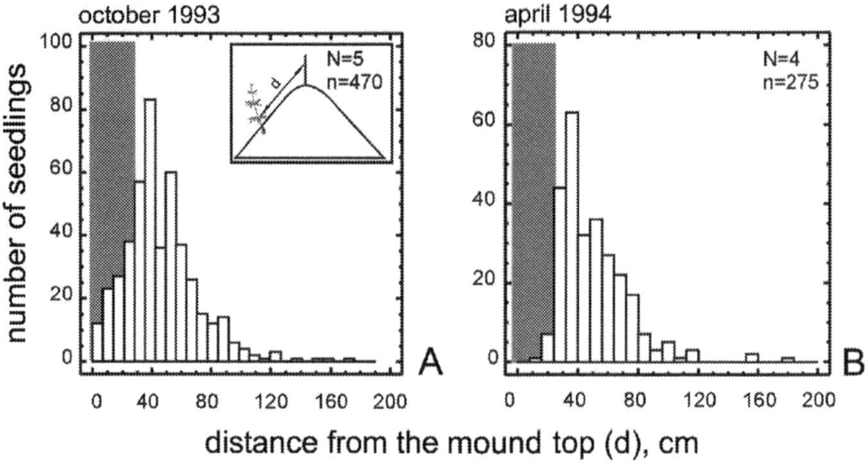

Fig. 10.3. Frequency histograms of seedling distances from the centre of the mound top in October (A) and in April (B). Grey columns indicate the area of highest mortality of seedlings.

From measurements taken at the same time, plants from ant mounds have significantly longer stems and a higher number of nodes than plants on surrounding sites (Table 10.1). This difference was the highest in October (Figs. 10.4, A; 10.5, A) and the lowest in April – May of the next year (Figs. 10.4, B; 10.5, B).

Table 10.1. Length and node number in *Galium aparine* plants from various microsites. Abbreviations: n, number of recordings. Average values are given in bold print, standard deviations – in italics.

Date	On ant mounds		Off ant mounds		n
	Length, cm	Nodes, number	Length, cm	Nodes, number	
October 1994	**9.40** *5.30*	**3.7** *2.2*	**5.70** *2.43*	**1.6** *1.0*	388
April 1995	**7.95** *3.97*	**5.5** *2.7*	**6.29** *2.09*	**3.7** *1.6*	367
June 1995	**9.40** *5.30*	**12.5** *2.6*	**51.35** *9.21*	**9.1** *1.6*	107

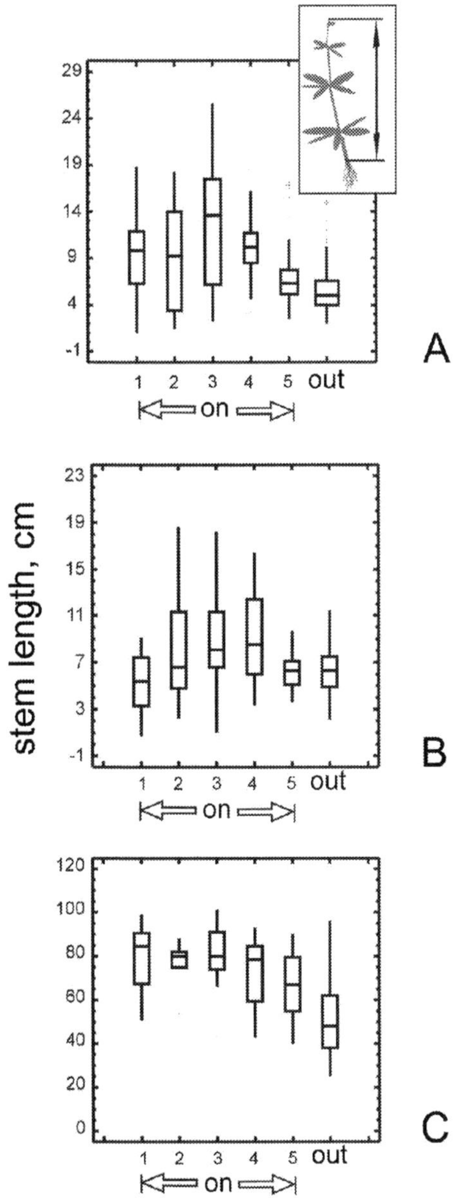

Fig. 10.4. Box-and-whiskers diagrams showing data on the length of seedlings (L) collected from different microsites in October (A), April (B), and June (C). on, plants from five ant mounds; out, control plants from five sites in the forest.

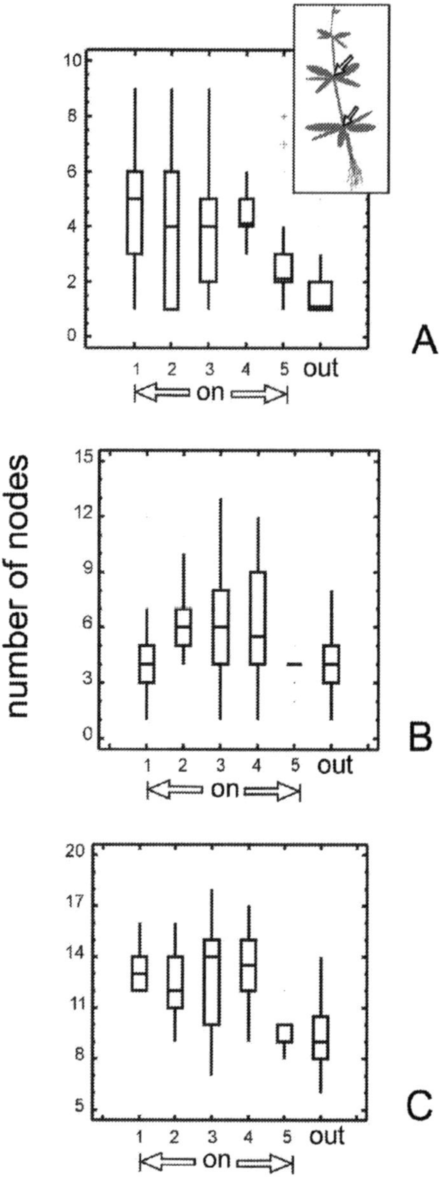

Fig. 10.5. Box-and-whiskers diagrams showing data on the number of nodes in seedlings collected from different microsites in October (A), April (B), and June (C). on, plants from five ant mounds; out, control plants from five sites in the forest.

2. ASYMMETRY OF INTERACTIONS BETWEEN NON-MYRMECOCHORES AND ANTS

2.1 Plant concentration on the nest mounds

Plants of some species tend to concentrate in the vicinity of ant mounds due to specific properties of the soil at these microsites. A complex of plant species connected to mounds of *Lasius flavus* ants has been previously reported (Woodel and King, 1991). Plant species composition around nests of *F. rufa* s. l. ants also differs from surrounding microsites (Forel, 1874). Many nitrophilous plant species, such a *Urtica dioica* (Grimalsky, 1960; Dlussky, 1967), *Rubus saxatilis* (Lebedev, 1949), *Rosa canina, Rubus idaeus* (Kostrin, 1951), concentrate around mounds of ants from the genus *Formica*. Some of these plants produce a considerably higher amount of fruits, while growing on the mounds or in its vicinity (Kostrin, 1951). Thus, even non-myrmecochores seem to have some advantages from growth on ant mounds. Diaspore transport by ants has not been previously observed in these species.

2.2 Why do ants remove and transport diaspores without elaiosomes?

Formica polyctena ants build large nest mounds from dry parts of plants (0.3 – 2.5 mm in diameter). Ant workers collect the building material and transport it to the nest. Since the size of *G. aparine* diaspores corresponds to size of pieces preferably used by *F. polyctena* workers as building material, removal and transport of diaspores in this plant species may be explained merely by the use of the diaspores as building material. Also, cleaning of the foraging routes may lead to diaspore relocation. However, Sernander (1906) has reported that *Lasius niger* ants, building soil mounds and not using pieces of plants for this purpose, also removed *G. aparine* diaspores rather actively. Time of half removal was about 20 min (n=1). This result is difficult to explain without further experiments.

Removal rates of *G. aparine* diaspores by *F. polyctena* workers was 50 – 100 times lower than for elaiosome-bearing diaspores of myrmecochores (Gorb and Gorb, 1995a; b; 2001a; Gorb, 1998). However, *G. aparine* diaspores may reach ant nests due to transport by ants. The presence of the high amount of diaspores in soil samples from ant mounds may be explained by such ant behaviour. Additionally, diaspores of plants growing on nests remain in nests and concentrate there, because of the absence of the secondary relocation for seeds without elaiosomes.

2.3 Character of interactions between non-myrmecochores and ants

Plant abundance on ant mounds mainly depends on the properties of the nest substrate (Radchenko et al., 1979). Nest substrate is more nutritious than the surrounding soil and has some specific physical and chemical characteristics (Greenfeld, 1941; Malozemova and Koruma, 1973; Czervinski et al., 1971; King, 1977a; b; c; Petal, 1978; Davidson and Morton, 1981a; Beattie and Culver, 1983; Culver and Beattie, 1983). Our data show that mound conditions of the *F. polyctena* ants are favourable for diaspore germination and seedling growth of *G. aparine*, especially at early stages of development. Presumably, the higher humidity of the mound substrate in August promotes an early seed germination. High nutritional content of the substrate provides fast growth of seedlings on nest mounds.

In October, when seedlings cover the mound and shade it, some of the plants perish. One may suggest that ants kill plants to free the mound surface. In the Spring, ant activity is much higher and they begin to kill plants by covering them with mound substrate on the mound top and partly on the slopes. This is the reason why seedlings on the mound top perish during this period. In the late Spring, individuals of *G. aparine* that survived on the mound, again start to shade the mound. However, ants are not able to kill mature plants which escaped covering due to the fast growth. In the forest, abandoned *F. polyctena* nests completely covered by *G. aparine* plants can be observed.

In contrast to the mutualism between ants and myrmecochores, interactions between non-myrmecochorous *G. aparine* and *F. polyctena* ants are asymmetrical (Gorb et al., 1997; Sindarovskaya et al., 1997). Ants provide some advantages for the plant. First, they are additional dispersal agents of diaspores. Since poor soil may often be a limiting factor for plant growth, nutrient-enriched mound substrate may promote seed germination and seedling growth. On the other hand, a very high density of seedlings on nests and then killing by ants lead to the high mortality of plants located on nests. Unlike myrmecochores with elaiosome-bearing diaspores, *G. aparine* does not provide ants with energetically-valuable food. Moreover, a large accumulation of seedlings and mature plants on the nest mound shade the nest and, probably, leads to a decrease in the nest temperature. Such an effect may be critical for the colony and may result in the colony leaving the nest. Killing plants and colony relocation are time-consuming events and disadvantageous for the ant colony.

According to the literature, interactions of the type «host-visitor» are not always mutualistic (Cushman and Beattie, 1991). It seems that interactions between the non-myrmecochore *G. aparine* and *F. polyctena* ants are incidental compared to the true mutualism between myrmecochores and ants. However, this example of such a type of interaction between ants and plants may aid in the understanding of the relationship between costs and benefits for both participants of such an accidental connection. These interactions could be considered as potential precursor interactions prior to a closer interaction between plants and ants and, depending on the future evolutionary contingencies, lead to myrmecochory.

3. SUMMARY

- Non-myrmecochorous plant species are able to colonise nest mounds. In this chapter, the results of the experimental study of such interactions are presented. Removal rates of fruits of non-myrmecochore *Galium aparine* by *Formica polyctena* ants were studied. Soil seed material, seedling patterns and growth dynamics of plants from the nests versus those from the forest were compared.
- Removal rates of fruits by ant workers are very low. However, the soil samples from nests contain a higher number of diaspores than those from randomly chosen sites in the forest.
- Diaspores start to germinate on the nest mounds significantly earlier than at surrounding places. The seedling density was higher within experimental squares located on mounds. Plants from the nests are significantly longer, with a higher number of nodes and a higher mortality rate than at surrounding sites.
- In spite of the higher mortality rate of seedlings on mounds, ants provide some benefits to the plants. First, ant workers are additional dispersal agents for *G. aparine* fruits. Furthermore, the soil properties of ant mounds are beneficial for seed germination and seedling growth.
- At the same time, *G. aparine* does not provide ants with energetically valuable food as do myrmecochores with elaiosome-bearing seeds. Moreover, seedlings and plants on the mounds shade the nests.
- The interactions between *G. aparine* and *F. polyctena* ants are asymmetrical. The interactions are more beneficial for the plants than for the ants.

Chapter 11

Methods for studying myrmecochory

The previous chapters demonstrated some of the plant adaptations to myrmecochory, the principal stages of diaspore dispersal by ants, and the influence of various factors on the effectiveness of dispersal in different stages as well as advantages obtained by plants from this method of dispersal. Some data were taken from the literature, but most are based on the original experiments by the us. This chapter describes the objects and methods of these experimental studies. Several methods have been used by previous authors and are standard for resolving particular questions on diaspore dispersal by ants. Some methods were developed during the course of this study. Information given in the chapter may be used as a kind of manual for those who wish to study plant dispersal by ants. In the first half of the chapter, study site, species of plants and ants, and morphological and anatomical methods are described. In the second half, experimental methods are reported according to their appearance in previous chapters.

1. STUDY SITE

The majority of experiments and observations were carried out in the oak-maple deciduous forest near Jagotyn (Kyiv province, Ukraine) (Fig. 11.1; Appendix, Fig. IX, A). Tree species such as *Quercus robur* L., *Acer platanoides* L., and *Ulmus laevis* Pall. are dominant. Dominant herbaceous plants include *Galium aparine* L., *Anthriscus sylvestris* (L.) Hoffm., *Asarum europaeum* L., *Lamium maculatum* L., *Viola hirta* L., *V. odorata* L., and *Glechoma hirsuta* Waldst. et Kit. Among the herbaceous species, a high number of myrmecochorous species occur, such as *Asarum europaeum*, *Ballota nigra* L., *Chelidonium majus* L., *Corydalis cava* (L.) Schweigg. et

Koerte, *C. bulbosa* (L.) DC., *Ficaria verna* Huds., *Gagea lutea* (L.) Ker-Gawl., *G. minima* (L.) Ker-Gawl., *Mercurialis perennis* L., *Omphalodes scorpioides* (Haenke) Schrank, *Pulmonaria obscura* Dumort., *Scilla bifolia* L., *S. sibirica* Haw., *Strophiostoma sparsiflora* (Mikan) Turcz., *Viola hirta* L., *V. matutina* Klok., *V. mirabilis* L., and *Viola odorata* L.

The effects of diaspore aggregation on the removal rates were studied in the oak-hornbeam deciduous forest of the Kaniv Natural Reserve (Cherkassy province, Ukraine) (Fig. 11.1). There, *Carpinus betulus* L., *Quercus robur* L., *Tilia cordata* Mill., *Acer platanoides* L., *A.negundo* L., and *Populus tremula* L. are dominant trees. *Equisetum sylvatica* L., *Aegopodium podagraria* L., *Carex pilosa* Scop., *Stellaria holostea* L., *Poa nemoralis* L., *Asarum europaeum* L., *Viola odorata* L., *Glechoma hirsuta* Waldst. et Kit. are dominant herbaceous plants.

Fig. 11.1. Map of the Ukraine showing locations of study sites.

2. PLANT SPECIES COVERED BY THE EXPERIMENTS

Experiments were carried out with plant species growing at the study site with diaspores, which were accessible to ants (Appendix, Fig. XIV; XV). According to the diaspore dispersal methods, species studied belong to three main groups: (1) obligate myrmecochores, (2) facultative myrmecochores, and (3) non-myrmecochores. The first group is composed of species, whose diaspores are dispersed only by ants. Stems and pedicles may decline and lie down. Some other plants demonstrate *geocarpy*, when fruits and seeds ripen in the soil, or *autobarochory*, when diaspores fall off due to gravity, and are further dispersed by ants (terminology after Levina (1957; 1967)). Facultative myrmecochores possess other methods of diaspore dispersal, such as different variants of *zoochory*, *autochory* (*automechanochory*), *anemochory*, and *hydrochory*, in addition to myrmecochory. Non-myrmecochores are plant species which do not possess any specialised adaptations for diaspore dispersal by ants and are usually dispersed by other dispersal methods. Below, a brief information about plant species used in the field experiments is given.

2.1 Obligate myrmecochores

***Asarum europaeum* L.** (Aristolochiaceae) is a shade-requiring clone-generating perennial plant with short creeping shoots. Fruiting is in June – July. Fruits are round capsules located at the soil surface or slightly submerged in soil. Fruits contain 10 – 30 seeds and dehisce after ripening. Seed producing capacity ranges from 15 to 50 seeds per plant. After fruit dehiscence, seeds remain in the group until they are found by ants or other animals.

***Corydalis cava* (L.)** Schweigg. et Koerte (Fumariaceae) is a tuber-bearing perennial plant. The fruiting is in May. Fruits are dry pod-like capsules, in which both leaves open. After fruit ripening, pedicles decline to the soil surface. Seed producing capacity ranges from 30 to 60 seeds per plant. Seeds are usually accessible in groups to ants.

***C. bulbosa* (L.) DC.** (Fumariaceae) is also a tuber-bearing perennial plant bearing fruits in late April – early May. After fruit ripening, pedicles decline to the soil surface. Seed producing capacity ranges from 50 to 100 seeds per plant. After fruit dehiscence, seeds are usually accessible in groups to ants.

***Pulmonaria obscura* Dumort.** (Boraginaceae) is a perennial plant with short rhizomes. The fruiting is in late May – early June. During diaspore ripening, pedicles may decline to the soil surface. The fruit is coenobium in an effused flower bell. It is able to split into separate parts called erems. Seed producing capacity ranges from 15 to 30 erems per plant. Erems are mostly scattered on the soil or litter surface, but may also be accessible in groups to foraging ant individuals.

***Viola hirta* L.** (Violaceae) is a shade-requiring clone-generating perennial plant with short creeping shoots. The fruiting is from June to September. Fruits are round capsules located on the soil surface or slightly submerged in the soil. Ripe capsules contain 10 – 30 seeds. Seed producing capacity ranges from 15 to 70 seeds per plant. After fruit dehiscence, seeds remain in the group until they are found by ants or other animals.

***V. odorata* L.** (Violaceae) is a clone-generating perennial plant with short creeping shoots. The fruiting is from June to September. Fruits are round capsules containing 10 – 30 seeds per fruit. Seed producing capacity ranges from 15 to 70 seeds per plant. The majority of fruits originate from cleistogamous flowers. Fruits from chasmogamous flowers are located on the litter surface, whereas fruits from cleistogamous flowers are in the litter or may be slightly submerged in the soil. After fruit dehiscence, seeds remain in the group.

2.2 Facultative myrmecochores

***Chelidonium majus* L.** (Papaveraceae) is a herbaceous perennial plant with the branching stem. Fruiting is from May to October. Fruits are dry pod-like capsules containing 50 – 100 seeds per fruit. Seed producing capacity ranges from 1000 to 5000 seeds per plant. Seed dispersal contains two phases, autochory and myrmecochory. After the first dispersal stage, seeds are scattered on/under the litter surface and are solitarily accessible to ants.

***V. matutina* Klok.** (Violaceae) is an annual or biennial light-requiring plant with an ascending stem. The fruiting is from May to October. Fruits are triangular capsules with 20 – 70 seeds. Seed producing capacity ranges from 200 to 1000 seeds per plant. After fruit dehiscence, seeds are ejected from the fruit and scattered on/under the litter surface.

***V. mirabilis* L.** (Violaceae) is a shade-requiring perennial plant with upright stems. Most of fruits originate from cleistogamous flowers in the period from June to September. Fruits are triangular capsules with 15 – 30 seeds ejecting from the fruit and scattering on/under the litter surface after fruit ripening. Seed producing capacity ranges from 50 to 100 seeds per plant.

2.3 Non-myrmecochores

***Galium aparine* L.** (Rubiaceae) is an annual herbaceous climber. The fruiting is from July to October. After ripening, the fruit splits into two round mericarps, with little hooks on their surfaces. These fruits are highly specialised for *epizoochory*.

3. ANT SPECIES USED IN THE EXPERIMENTS

In the temperate deciduous forests, one of the main diaspore dispersal agents for myrmecochorous diaspores are ants from the *Formica rufa* s.l. group. A considerable part of the food ration of *F. polyctena* Foerst. (Formicinae) ants are insects and other small arthropods (Dlussky, 1967). In the late Spring and early Summer, these ants are one of the main consumers of elaiosomes. *F. polyctena* colonies occur mostly in well-illumined sites with a low density of trees. These ants build neat nest mounds out of small plant fragments. The density of *F. polyctena* colonies at the study site was about 300 colonies per 1 km^2. Another formicine species, *Lasius fuliginosus* Latr., was quite rare. Two myrmicine species, *Myrmica rubra* L. and *Leptothorax nylanderi* Foerst. were common.

Workers of *F. polyctena* are relatively large (body length 5.0 – 7.5 mm). *L. nylanderi* are very small (2.2 – 2.7 mm). *L. fuliginosus* and *M. rubra* are of an intermediate size (4.5 – 5.0 mm). Diet and foraging behaviour of the four species are species-specific. Workers of *F. polyctena, M. rubra*, and *L. nylanderi* are predominately predators: they collect mainly insects and sometimes feed on aphid honey-dew. *L. fuliginosus* mainly feed on the aphid honey-dew and collect insects and elaiosome-bearing seeds only as a supplementary kind of food.

F. polyctena workers usually forage on the litter surface, whereas *M. rubra, L. fuliginosus*, and *L. nylanderi* – on and in the litter as well. *F. polyctena* and *L. fuliginosus* possess relatively large foraging territories (20 – 60 m in diameter), which do not overlap with each other. *M. rubra* and *L. nylanderi* territories are considerably smaller (up to several meters in diameter). Territories of the two latter species may be located within territories of the first two species.

4. METHODS

To study various aspects of seed dispersal by ants, morphological, anatomical, phenological methods were applied. Additionally, field observations and experiments were carried out. To understand seed flow on ant territories, the soil seed bank from various microsites of the forest was studied. Results obtained were analysed by a complex of statistical methods.

4.1 Morphology and anatomy

For **morphological** studies, fresh ripe diaspores were used. A binocular microscope was used for drawing and size measurements of diaspores and elaiosomes. For *anatomical* investigations, myrmecochorous diaspores were fixed in 70% ethanol and dehydrated in an increasing series of ethanol and propylene oxide, and embedded in a low viscosity epoxy resin Spurr (Serva) (Spurr, 1969). Sections of 10 – 15 µm were made in equatorial and medial planes, stained in toluidine blue, mounted on the glass slides in epoxy resin Depex (Serva), and studied with a Zeiss Axioplane light microscope.

To study the **distribution of lipid-like substances in elaiosomes**, frozen sections (10 – 15 µm thick) were prepared from diaspores previously fixed in 3% formaldehyde. Sections were mounted on glass slides, rinsed briefly in distilled water, washed for about three minutes in 70% ethanol, stained 15 – 30 min with an ethanol solution of the Sudan Black, rinsed in 50% ethanol, mounted in glycerine-gelatine, covered with cover-slips, and studied with the light microscope.

Fatty vacuoles in the elaiosome cells were visualised by the following technique. Diaspores were fixed in glutaraldehyde in phosphate buffer (pH=7.3) for 12 h at a temperature of 4 °C, then postfixed in a 1% solution of osmium tetroxide (OsO_4) in phosphate buffer. The samples were dehydrated in an ascending row of ethanol and propylene oxide, and embedded in a low viscosity epoxy resin Spurr. After polymerisation, sections were cut from the resin blocks using a diamond knife. Semi-thin sections (1 – 2 µm) were mounted on round cover-slips having a diameter of 1 cm, rinsed for 1 – 2 min in Maxwell solution (Maxwell, 1978) in order to remove epoxy resin. Samples were then rinsed in absolute ethanol, critical-point dried, sputter coated with gold-palladium (10 nm), and observed in a scanning electron microscope Hitachi S-800.

4.2 Field observations and experiments

4.2.1 Studies on the myrmecochorous syndrome

To determine relationships between fruiting periods in plants and periods of ant activity, data on fruiting phenology of plants were collected for myrmecochores during four successive seasons (1993 – 1996).

Mechanisms of diaspore deposition into sites accessible to ant foragers were carried out with *V. odorata* seeds and a number of ant species. In this species, fruits are deposited on the litter or soil surface but sometimes buried into the litter. Experiments were carried out on warm sunny days from 11 a.m. to 5 p.m. within the colony territory of *F. polyctena*, where workers of this ant species tend to forage mainly on the litter surface. Other ant species, such as *M. rubra* and *L. nylanderi,* forage on and in the litter.

A glass plate (2x10 cm) was placed on the litter surface in the vicinity of about 5 m from the *F. polyctena* nest and at a considerable distance from foraging routes of this species. Depots containing 10 seeds each were deposited into small plastic containers (1 cm in diameter), attached above or below the glass plate (Fig. 11.2). Such an arrangement of seed depots allowed seed deposition on and in the litter and made possible a simultaneous observation of both containers by the experimenter. Ant species and ant action (seed removal or refusal) were recorded for each ant visit. Prior to the beginning of each new experiment, plate location was changed to prevent effects of learning and recruitment on the seed removal. In all, 15 repetitions of the experiment were carried out.

4.2.2 «Cafeteria» experiments

«Cafeteria» experiments were carried out (about 200 h total) to evaluate diaspore removal rates by ants. Seeds of five plant species (*A. europaeum, V. hirta, V. matutina, V. mirabilis, Ch. majus*), setting fruits in June – August, and *F. polyctena* ants were used in the experiments. Seeds for experiments were collected at sites located in the vicinity of the study site, but not directly in the study site. Collected seeds were kept in the refrigerator and not used for more than 2 – 3 days. Five types of «cafeteria» experiments were performed.

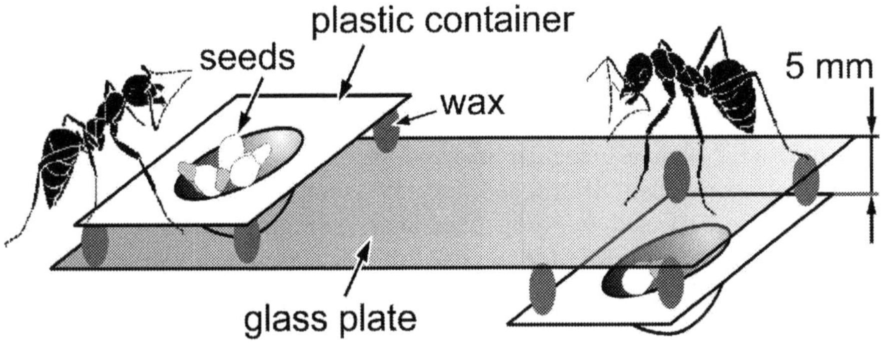

Fig. 11.2. Set-up of the experiment studying the effect of seed location on removal rates by ants.

(1) To evaluate seed removal rates by *F. polyctena* ants, 10 seeds of a single plant species were deposited on a small wooden table (6x6 cm) (Appendix, Fig. IX, F) in the vicinity of 5 – 10 cm from the foraging route and at a distance of 5 m from the *F. polyctena* nest, where mainly foragers of

the species would remove seeds. Seed removal was detected visually and any change in the number of seeds was recorded. Data obtained were averaged for each plant species (7 – 22 repetitions for each species).

(2) Seed removal rates by the complex of ant species occurring at the site in the forest were evaluated as follows. Depots containing 10 seeds of each plant species were deposited at five randomly selected microsites at a distance of 15 m from *F. polyctena* nests, where ants of other species were able to visit and remove seeds. The number of seeds remaining in the depots were counted for each 30 min of exposure. For each plant species, four repetitions of the experiment were performed. Results obtained were also compared with the results obtained in experiments carried out in the vicinity of foraging routes of *F. polyctena.*

(3) To evaluate ant preferences in the removal of seeds of various plant species, experiments with paired presentation of plant species were carried out. 10+10 seeds of two plant species were deposited as a single depot in the vicinity of the ant route. The experiment was then performed as described in (1). 10 possible pairs for five plant species were tested.

(4) To evaluate ant preferences in selection of seeds of the same plant species, removal rates of seeds with partially (50%) ablated elaiosome or seed bodies were studied. Removal rates of intact seeds were compared with those of seeds with partially ablated seed bodies or elaiosomes. 10 intact and 10 treated seeds were deposited as a depot in the vicinity of the ant route. The experiment was then performed as described in (1).

(5) To evaluate differences in the size of ant workers removing seeds of various plant species, removal of seeds by *F. polyctena* foragers of different size classes was studied. The experiment was carried out as described above (1). After seed removal, each ant individual was captured and its head width was measured with a binocular microscope. Size classes of *F. polyctena* individuals were also determined for the colony. For this purpose, ant workers were collected from the litter at the foraging routes. Data on the head width were compared for samples of ants removing seeds of different plant species and for ant samples from the forest litter.

4.2.3 Factors influencing diaspore removal rate

To study the **effects of diaspore size** on the removal rate by *F. polyctena* ants, «cafeteria» experiments were performed in April – May with seeds of two closely-related plant species, *C. cava* and *C. bulbosa,* which occupy the same microhabitats in the forest. A similar procedure as described in 4.4.2 (1) was applied. 22 – 24 repetitions were carried out for each plant species. To compare size classes of ant individuals, giving their preferences to seeds of particular plant species, the experiments were performed as described in 4.4.2 (5).

The **role of the presence of the elaiosome, diaspore body, and their contents** in attracting of *F. polyctena* ants was studied for two plant species, *C. cava* and *P. obscura.* Two types of experiments were carried out on territories of two colonies of *F. polyctena.*

(1) Elaiosomes were separated from fresh diaspores by tweezers and/or sharp razor blade. Diaspore bodies and the elaiosomes were used for the preparation of juices which were pressed out from diaspore bodies and elaiosomes. Round balls 2 – 3 mm in diameter were made out of wet filter paper. These were air dried and used as a substrate for juice infiltration (about 0.01 mg of solution per ball). 10 balls infiltrated with the juice of the diaspore body or elaiosome were exposed to ants. Balls infiltrated with distilled water and intact diaspores of both plant species were taken as control objects.

(2) In the second experiment, removal rates of diaspores with ablated elaiosomes were compared to intact diaspores of each plant species.

In both experiments, seed depots were exposed to ants at a distance of 15 – 20 cm from the foraging route and at a distance of about 3 m from the *F. polyctena* nest. Changes in the number of balls or diaspores of experimental depots were recorded versus exposure time. 5 – 20 repetitions with each type of ball and diaspore were carried out. To prevent influence of learning and recruitment of ant workers, locations of tables were changed after each repetition.

To study the **effects of diaspore aggregation on removal rates by F. polyctena ants**, «cafeteria» experiments were performed in June – July with seeds of two plant species, *Ch. majus* and *V. odorata. V. odorata* seeds are usually accessible in groups to ants, whereas seeds of *Ch. majus* are scattered after the autochorous phase of dispersal and accessible to ants as solitary objects.

Seeds were exposed to ants on wooden tables (10×10 cm) with 10 small depressions located at some distance from each other. Tables with six different distances between depressions (0, 1, 2, 4, 8, 16 mm) were used (Fig. 11.3). Diaspores were deposited into depressions to prevent their occasional displacement by ants. All six tables with 10 diaspores on each were simultaneously used in the experiment. Tables were placed at a distance of 20 – 30 cm from each other, close (20 cm) to the foraging route at a distance of about 5 m from the nest. Each seed removal was recorded versus exposure time. In all, 21 repetitions were carried out with each plant species. The order of the table position was changed after each repetition.

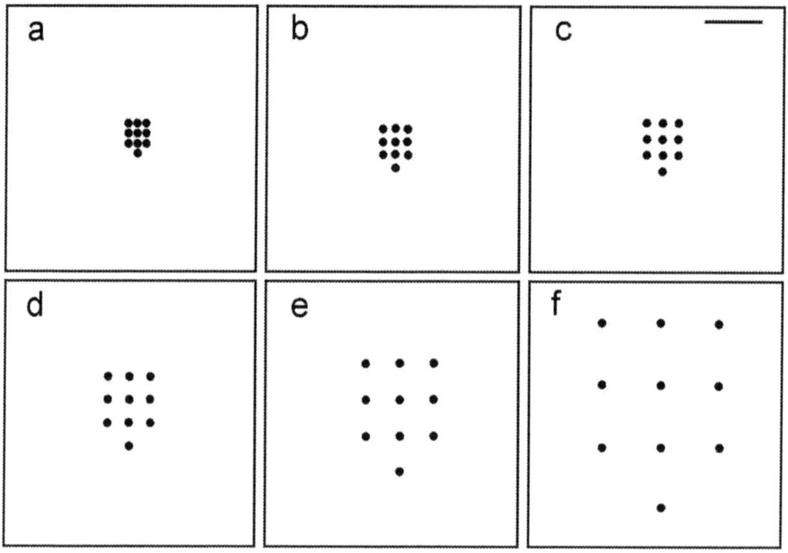

Fig. 11.3. The six patterns of depressions on wooden table that were used in the seed-removal experiments. The distances between neighbouring depressions were (a) 0, (b) 1, (c) 2, (d) 4, (e) 8, and (f) 16 mm. Scale in top right corner: 20 mm.

4.2.4 Effect of the ant species composition on diaspore removal

Seed removal of *V. odorata* seeds was studied in three microsites differing in the composition of ant species. The study was performed in June – August in an area of deciduous forest (200×200 м) containing three *F. polyctena* nests, one *L. fuliginosus* nest and numerous scattered nests of *M. rubra* and *L. nylanderi*. Within this area, three microsites with different ant species compositions were selected (*F. polyctena* territory: complex *F. polyctena* + *M. rubra* + *L. nylanderi*; *Lasius fuliginosus* territory: complex *L. fuliginosus* + *M. rubra* + *L. nylanderi*; *M. rubra* territory: complex *M. rubra* + *L. nylanderi*) (Fig. 11.4). This microsite selection allowed us to compare the total removal rate of diaspores in these microsites and contribution of workers of each ant species to the removal. Three types of experiments were carried out.

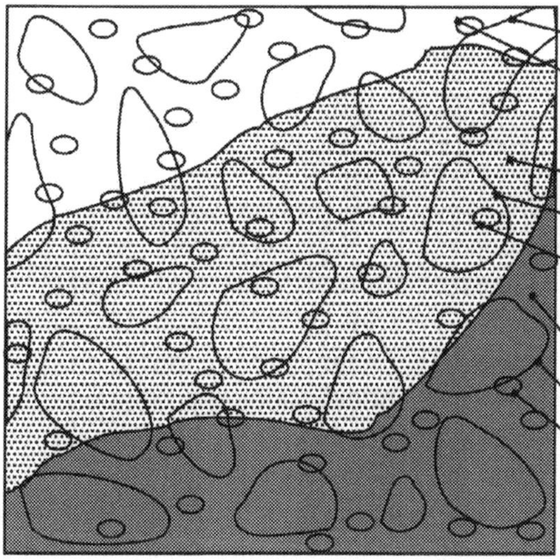

Fig. 11.4. A simplified diagram of microsites studied. FP, *Formica polyctena* territory; LF, *Lasius fuliginosus* territory; LN, *Leptothorax nylanderi* territory; MR, *Myrmica rubra* territory; *1*, MR+LN microsite; *2*, LF+MR+LN microsite; *3*, FP+MR+LN microsite.

(1) In the first experiment, the contribution of various ant species to the visiting rate of seed depots and to the removal rate of seeds was evaluated for various microsites. Initially, the seed predation rate by rodents, birds, and large ground beetles was evaluated. One half of the seed depots were covered by grid boxes with a mesh diameter of 7 mm, which allowed ants to visit depots and prevented larger animals from doing so. The remaining depots were exposed to any kind of visitors. Since differences between covered and uncovered depots did not vary significantly, only uncovered depots were used in later experiments.

In each of three microsites, 10 wooden tables (6×6 cm) with depots of 20 seed each were exposed to ants at randomly selected places. Depots were visited every 30 min over 4 h, and number of remaining seeds was counted, after which the depot was replenished to the amount of 20 seeds. In all, four repetitions of the experiment were carried out on each microsite.

(2) Ants usually consume a part of the elaiosome prior to seed removal. To evaluate the relationship between elaiosome consumption and seed removal rate, direct observations of ant behaviour at seed depots were made. In each microsite, wooden tables with depots of 20 seeds each were exposed to ants. Each depot was visually observed and the following information was recorded: 1) ant species; 2) arrival and departure time; 3) seed number tested

by the ant individual; 4) whether the seed was removed; 6) time of manipulation with each seed. If the seed was removed by an ant, one seed was added by the experimenter to complete the depot to the amount of 20 seeds. Data obtained were pooled and averaged for each ant species for each microsite.

(3) Mark-recapture experiments were applied to evaluate the effects of learning and recruitment in ants *F. polyctena, M. rubra,* and *L. fuliginosus* on seed removal. Depots of 20 seeds each were exposed to ants in each microsite. Ant individuals, arriving to the seed depot, were marked with a fluorescent water colour. The same information as described in (2) was recorded about each ant visit. Additionally, the presence of the marker was recorded. Data were compared for two groups of ant individuals arriving at depots: marked ones and unmarked ones. Marked foragers were «experienced», since they were able to find the seed depot again after visiting nest or after dropping seeds during transport. Unmarked foragers were «inexperienced», since they arrived at the depot for the first time.

4.2.5 Diaspore transport by ants

Methods of diaspore transport by ant workers of *F. polyctena, L. fuliginosus, M. rubra*, and *L. nylanderi* were determined visually after ants removed seeds from the depot. In the case of *F. polyctena* workers, transporting methods were compared for various seeds (*V. odorata, A. europaeum, V. hirta, V. mirabilis, V. matutina, Ch. majus*). Ant handling of seeds and transporting methods were photographed.

Since diaspore dropping during ant transport is an important parameter to evaluate efficiency of myrmecochory for particular plant species, seed transport by *F. polyctena* ants was studied in August with seeds of *A. europaeum, V. hirta, V. mirabilis, V. matutina,* and *Ch. majus* (about 50 h of field observations and experiments). Seeds were collected in the vicinity of the study site, but not directly in the study site. Collected seeds were kept in the refrigerator and used within 2 – 3 days. In all, 1020 interactions of ants with seeds were recorded. Three series of experiments were carried out.

(1) Mark-recapture experiments were carried out to evaluate seed dropping rate during transport by ants. To prevent occasional replacement of seeds in the depot, marked diaspores were used. A depot of 100 seeds was deposited on the foraging route at a distance of 4 m from the nest. Each ant worker that took a seed, was marked with fluorescent water colour. At a distance of 1 m from the ant nest, marked ants were recaptured, and both ants and transported seeds were removed from the experiment. Results were divided into three groups: 1) marked ants with marked seeds; 2) marked ants without seeds; 3) unmarked ants with marked seeds. Results were pooled

together from repetitions for each plant species, since data in each repetition did not differ considerably (n=100, N=3).

(2) Effect of ant size on the seed dropping rate was studied in the experiment. The head widths of those ant workers who reached a three meter distance with the seed, were measured with a binocular microscope. These data were compared with the head widths of ant individuals who removed seeds of the same plant species from the seed depot (see also 4.2.2 (5)). In all, head sizes of 483 ant individuals were measured in the experiment.

4.2.6 Plant diaspore flows on ant territories

Direct observations of seed flow on ant territories were made in May during the period of the most intensive gathering of elaiosome-bearing diaspores by ants. Since the majority of early-spring ephemeral myrmecochores set their fruits at this time, we were able to estimate maximal values of the diaspore flow. The area of the foraging route (20 cm wide) was selected at about 2 m away from the *F. polyctena* nest. Once a week, the number of diaspores, having crossed this area by ant transport, was counted. Plant species of the diaspores were also determined.

4.2.7 Secondary relocation of diaspores by ants

A square area (30×20 m) of the forest with three *F polyctena* nests and one *L. fuliginosus* nest was selected so that territories of colonies of both species had a joint border. In the experiment, the influence of each ant species on the spatial distribution of diaspores, mature plants, and seedlings of myrmecochores was estimated. The number of mature plants and seedlings was counted in 40 experimental squares (25×25 cm) randomly selected (1) on *F. polyctena* territory at a distance of 10 m from the nest (n=10); (2) on *L. fuliginosus* territory at a distance of 2 – 10 m from the nest (n=10); (3) on the old border between territories (n=10), and (4) on the new border between territories (n=10) (Fig. 11.5). Since the border between territories tends to change its shape depending on growth rates of both colonies, both old and new borders between territories were distinguished.

4.2.8 Ecological aspects of the myrmecochory

To estimate **advantages for *V. odorata* plants of growing on *F. polyctena* mounds**, plants from 10 ant nests and from 10 randomly selected control sites at a distance of 5 – 6 m from the nest were compared in July – August. For 376 plants (176 from ant nests and 200 from control sites), fruit number, fruit symmetry, depth of the fruit location in the soil, mound material, or forest litter were estimated. For 78 fruits (41 from ant nests and 37 from control sites), seed number per fruit, and fruit diameter in two directions were estimated.

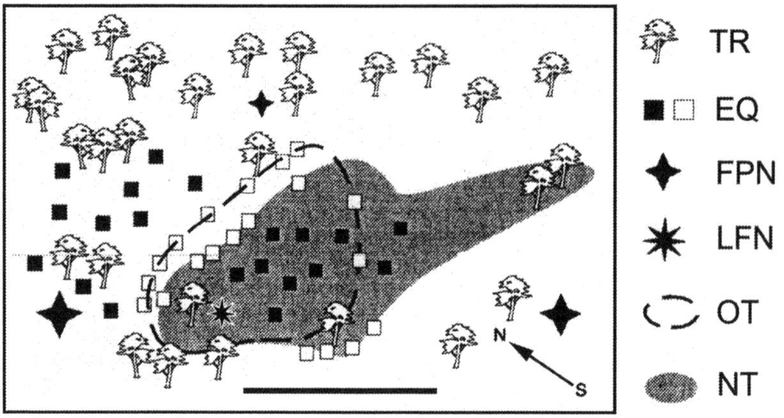

Fig. 11.5. Diagram of the study site. EQ, experimental squares (25×25 cm) in ant territories (black) and on borders (white); FPN, *Formica polyctena* nest; LFN, *Lasius fuliginosu* nest; NT, new territory of *L. fuliginosus*; OT, old territory of *L. fuliginosus*; TR, tree. Scale bar: 10 m.

Ant ability to damage ripe fruits of *V. odorata* was tested as follows. Freshly-collected fruits were exposed to ants directly on *F. polyctena* mounds. To prevent carrying away of whole fruits by ant workers, pedicles were tethered to a wooden rod attached firmly to the ground. To prevent quick drying and cracking of fruits, pedicles were wrapped in wet porous filter paper. Fruits exposed to ants were visited 24 h after exposure. Damage to fruits was recorded. Similar experiments were also carried out with other ant species (*Lasius flavus, L. fuliginosus, L. niger, M. rubra*).

In order to estimate such a selective advantage of the myrmecochory, as a **decrease of seedling density in the vicinity of the parent plant**, a two-year experiment was carried out with *C. bulbosa*. The experiment was set up on a site (3.8×2.6 m) containing 16 mature plants of *C. bulbosa*, which had previously been planted there two years before. To decrease the influence of other factors on the results, the study site was cleared of other plant species and seedlings of *C. bulbosa*. *Lasius niger* ants were the only ant species at the study site and therefore the only potential dispersal agents. To easily estimate orthogonal co-ordinates of plants, the study site was divided into 247 squares (20×20 cm) (Appendix, Fig. IX, E). Each square was virtually divided further into other four equal squares during computer data processing (in all, 988 squares 10×10 cm). In May 1995, 355 seedlings were individually marked with small labels made out of aluminium foil. Orthogonal co-ordinates of mature plants and seedlings and seed producing capacity of mature plants were determined. To prevent the appearance of new seedlings, all fruits were removed. Repeated determination of orthogonal co-ordinates of 146 seedlings that survived was carried out in May 1996.

Data were processed with the aid of the computer program written for calculation of 1) distance from each seedling to the closest mature plant, 2) distance from each seedling to the nearest seedling, 3) distance from each square of 10×10 cm, containing seedlings (or without seedlings) to the nearest mature plant. These values were calculated for data obtained in each season and separately for surviving and dead seedlings. In addition, the program aided in visualisation of spatial distribution of seedlings and mature plants and hypothetical vectors of seed dispersal from mature plants.

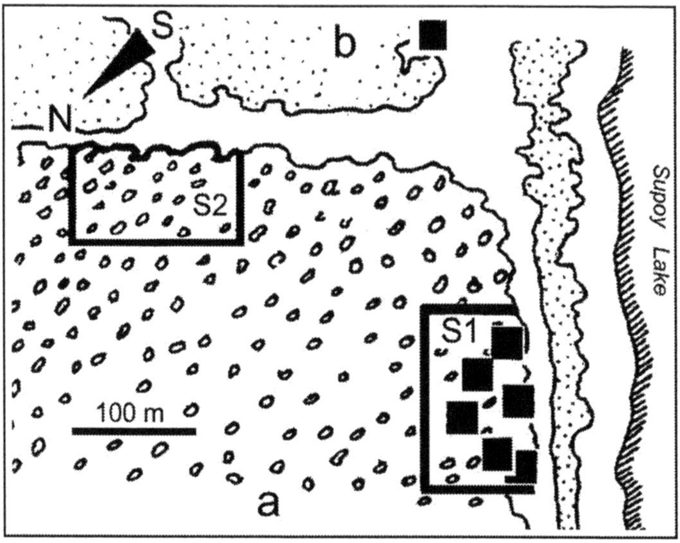

Fig. 11.6. Diagram of the study site of seedling mortality in *Asarum europaeum*. a, area with *A. europaeum* plants; b, area without the plants; S1, area with *Formica polyctena* nests; S2, area without the nests.

The **influence of myrmecochory on spatial distribution of seedlings and their mortality** was also studied in *A. europaeum*. Two experimental sites (150×100 м) were set up (Fig. 11.6). The first one contained six *F. polyctena* nests. The second one lacked nests of this ant species. Nests of other ant species, such as *L. nylanderi* and *M. rubra*, were present on both sites. 50 local clones of *A. europaeum* were selected on each site. The number of seedlings and juvenile plants with cotyledonous leaves, with one or two true leaves were counted in September 1993. Four circles of the seedling's location relative to the plant clone were recognised: C1 – inside the clone; C2 – outside of the clone, but inside the circle with a radius of 60 cm; C3 – within the ring with the outside and inside radii of 60 and 75 cm, respectively; C4 – within the ring with the outside and inside radii of 75 and 90 cm (Fig. 11.7). Seedling density on both sites was estimated by counting the seedling number in randomly selected squares of 1 m^2.

For estimation of the seedling mortality in this plant species, 100 seedlings that were located inside and outside of clones, were individually marked in November 1993. Repeated counting of the surviving seedlings was done in April 1994.

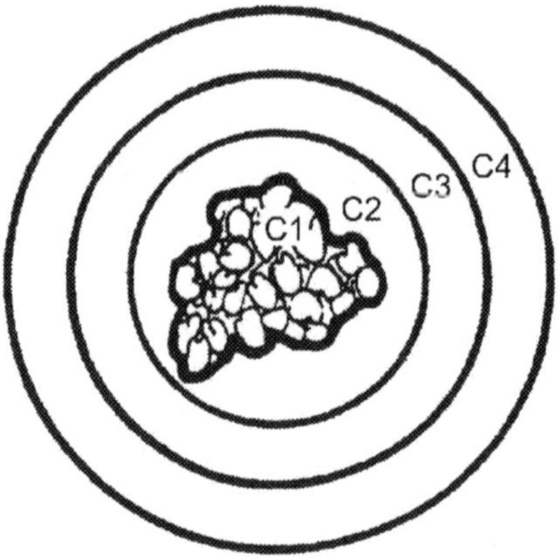

Fig. 11.7. Scheme of the seedling pattern in *Asarum europaeum*. C1 – C4, four levels of seedling occurrence; C1, in parental clone; C2, out of the clone in a circle with a 60 cm radius; C3, in ring with radii of 60 and 75 cm; C4, in ring with radii of 75 and 90 cm.

4.2.9 Interactions between ants and non-myrmecochorous plants

The non-myrmecochorous plant *G. aparine* and *F. polyctena* ants were used as model species in studies of interactions between ants and non-myrmecochorous plants. Three series of experiments were carried out.

(1) The ability of *F. polyctena* workers to remove mericarps (parts of fruits) of *G. aparine* was tested in a standard «cafeteria» experiment. 10 mericarps on a wooden table, located at a distance of 5 cm from the foraging

route and 4 m from the nest, were exposed to ants. To prevent mericarp predation by birds and rodents, tables were covered by metal grid boxes with a mesh of 7 mm. The tables were checked every 30 min to record changes in the diaspore number. In all, 12 repetitions of the experiment were analysed.

(2) Dynamics of both seedling density and growth were evaluated in a study site of 30×30 m containing 5 nests of *F. polyctena*. The seedling number was counted within 40 experimental squares (25×25 cm), randomly selected on the nest mounds (N=5, n=20) and outside of them at a distance of 4 – 8 m from ant nests (N=5, n=20). Records were carried out in the late Summer and Autumn 1993, and in the Spring 1994. For each seedling (n=740), the distance from the centre of the top of the mound to the seedling was measured.

(3) In plants from five ant mounds and five control microsites in the forest (n=862), the plant length and the numbers of nodes were estimated. The measurements were repeated in October 1994, April and June 1995.

4.3 Soil seed bank

Seed flow on the colony territory was estimated by the analysis of seed material of the nest substrate of *F. polyctena* ants compared to the soil seed banks of randomly selected control microsites in the forest. Soil samples were taken with a spatula from the superficial layers of soil (to a depth of 2 cm) from the foot of nest mounds (Fig. 11.8) (n=4) and from the control sites in the forest in the territory of the colony (n=5). The volume of each sample was approximately 1 dm^3. Samples were washed using a sieve with a mesh diameter of 1.2 mm and air dried for several hours. Diaspores were carefully taken from the soil samples with fine tweezers. Doubtful pieces were verified with a binocular microscope. In all, nine soil samples were processed and 8591 diaspores were determined.

For the **secondary relocation of diaspores by ants**, soil samples were collected from the foot of *F. polyctena* nests and in the vicinity of nest entrances of *L. fuliginosus* ants, from territories of both ant species, as well as from territory borders between both ant species. Soil was sampled to a depth of 2 – 4 cm and processed according to the method described above. In all, six soil samples were processed.

To estimate **seed flow of non-myrmecochores** within the ant territory, mericarps of *G. aparine* were collected from the soil samples taken from the foot of *F. polyctena* nest mounds (n=4) and from randomly selected sites of the colony territory (n=5). Sample processing was carried out according to the method described above. In all, nine soil samples were processed.

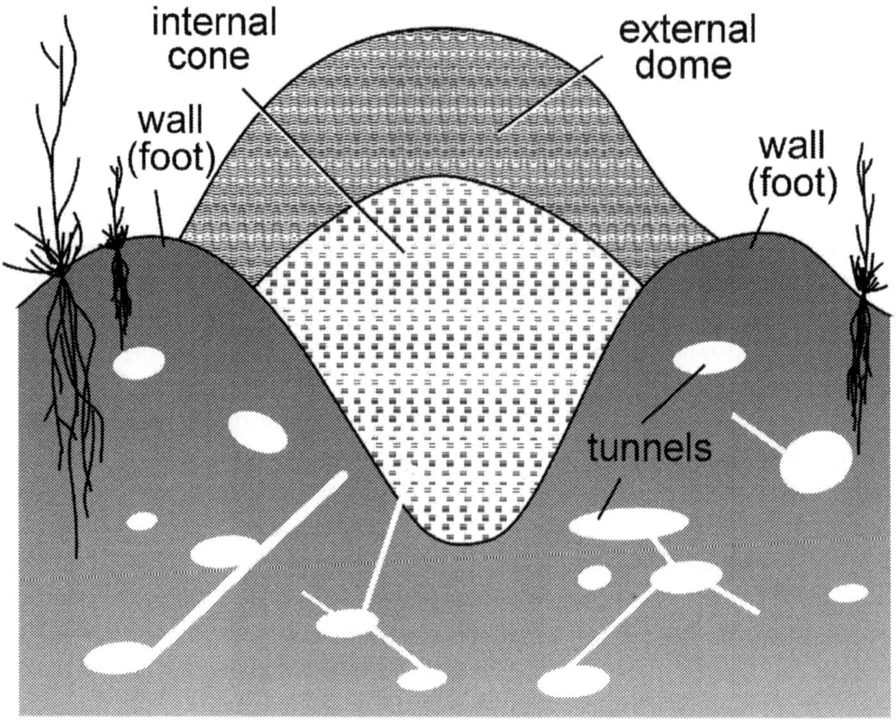

Fig. 11.8. Diagram of the nest of *Formica* s. str. (after Dlussky, 1967).

5. SUMMARY

- This chapter provides a short description of the study sites and experiments conducted by the authors.
- Common and original methods, such as morphological, anatomical, phenological, and experimental, as well as direct observations in the field and analysis of soil seed banks are described in detail.
- The following methods were used for the first time in studies of seed dispersal by ants: (1) the Maxwell method for the anatomy of diaspore appendages; (2) marking ants for evaluation of seed removal rates; (3) the analysis of seed soil banks; (4) modification of the «cafeteria» experiment for the study of an effect of seed deposition site on removal

rates; (5) the marking-recapture experiments for the evaluation of seed dropping rates during transport by ants; (6) the co-ordinate method of determination of seedling spatial distribution.

- Original computer simulations were created to analyse aspects of myrmecochory.

Conclusions and outlook

The general scenario of the diaspore dispersal by ants is rather clear: elaiosome-bearing diaspores are removed and transported by ants to their nests; ants consume energetically valuable elaiosomes and throw the diaspores away. Our results, however, demonstrate that current understanding of myrmecochorous relationships is too generalised. In this book, which is focused on one ecosystem with several spatially co-existing species of plants and ants, we show the range of possible mechanisms in plant-ant complexes. We quantified the ant attraction, diaspore transporting, dropping, and relocation; we explored geometric constraints on carrying abilities, examined factors causing seedling survival between microsites in relation to different ant behaviours, and presented computer simulations for showing how dropping and relocation can be influenced by different factors.

Myrmecochorous plants bear morphological, anatomical, biochemical and phenological features connected to ant dispersal. The complex of plant adaptations to this dispersal method is called myrmecochorous syndrome. The principal feature of the myrmecochorous syndrome is an elaiosome attached to the diaspore. Different myrmecochores have elaiosomes of diverse shapes, colours, sizes, anatomy, and origins. Different parts of ovule, pericarp or even flower and inflorescence may be initial tissues for the elaiosome origin. Elaiosomes contain a particular complex of fatty acids (mainly 1,2-diolein) which function as triggers for the carrying behaviour of ant workers. We show that chemical substances attracting ants may also be located in other parts of the diaspores. However, role of chemical cues in interactions between ants and myrmecochorous plants remains poorly understood. Also, studies on evolution of chemical substances releasing ant carrying behaviour may aid in understanding the multiple independent origin of myrmecochory in various lineages of flowering plants.

The first stage of myrmecochory is the removal of diaspores by ants. Diaspores of different plant species are removed at different rates. The exponential character of removal rates is especially pronounced for the obligate myrmecochores. Diaspore attractiveness increases with an increasing size of the diaspore.

Diaspores of myrmocochorous plants are usually removed by a variety of ant species contributing differently to the removal of the diaspores. Microhabitats in the deciduous forest differ in the composition of ant species. These differences may influence the dispersal of myrmecochorous plants. Ant competition for diaspores may result in a decrease in the consumption of elaiosomes during transport. Such features of ant behaviour as learning and recruitment may strongly effect the diaspore removal rates. This gives the diaspores a better chance of being quickly removed and thus prevents the drying out of the elaiosomes. Additionally, it reduces exposure of dispores to potential predators.

Ants carry elaiosome-bearing diaspores to their nests. Some of the diaspores reach the nest and are concentrated there. Others may be dropped by ants during transport. The dropped diaspores enter the soil seed bank. However, they might be repeatedly removed by other ant individuals and again carried in the direction of the nest. The number of diaspores reaching nests depends on dropping rates: diaspores with lower dropping rates have a higher chance of being deposited in nests. Facultative myrmecochores usually produce many small diaspores, which are characterised by low removal rates and high dropping rates during transport. Obligate myrmecochores produce rather few large diaspores, which have high removal rates and low dropping rates. An increase in the producing capacity of diaspores and a decrease in diaspore dropping rate result in an increase of maximal dispersal distance.

Ants can influence direction of the diaspore flow within the territory of an ant colony and the concentration of diaspores in nests. After removal of elaiosomes, diaspores of myrmecochores are transported by foragers of some ant species to territory borders (secondary relocation). Secondary relocation of diaspores to territory borders may result in an increase in the dispersal distances from the parent plant in both small- and large-seeded myrmecochores. In large-seeded myrmecochores this also results in a higher likelihood of reaching the «garbage dumps», which are nutrient enriched habitats. Myrmecochores may have additional benefits in the territories of ant species which relocate waste material to the territory borders. Computer simulation predicts that the dispersal system with secondary diaspore relocation provides a greater distance between diaspores and the parent plant as well as between diaspores, compared with systems without diaspore relocation. Such mechanisms may influence the plant species composition in various microsites of deciduous forest and formation of guilds of particular plant species.

Several case studies, carried out with various plant species, demonstrated advantages of myrmecochory for these species. Dispersal of diaspores for some distance from the parent plant reduces negative influence of the parent plant on seedlings, decreases seedling density, and, by this, decreases seedling mortality. Our studies demonstrate that selective advantages of diaspore dispersal by ants strongly depend on the ant species composition in the ecosystem and may be different for different plant species. That is why costs and benefits of myrmecochory can not be generalised for any plant species and should be defined for each particular plant species in a particular ecosystem. An important task for the future work remains to evaluate the relative importance of all hypothetical selective advantages of myrmecochory at least for one plant species.

Diaspores of non-myrmecochores with size classes corresponding to those of nest building material may be transported to nests due to the nest-building behaviour of ants. Diaspores of such plants, originating from diaspores which were brought into nests by ants, remain in the nests and are not relocated to the sites of the waste deposition. This mechanism leads to the colonisation of the nest mounds by non-myrmecochorous plant species. In spite of the higher mortality rate of seedlings on mounds, ants provide some benefits to the plants. First, ant workers are additional dispersal agents for diaspores of these plants. Furthermore, the soil properties of ant mounds are beneficial for diaspore germination of many plant species and their seedlings' growth. At the same time, these species do not provide ants with energetically valuable food as do myrmecochores with elaiosome-bearing diaspores. Moreover, seedlings and plants on the mounds shade the nests. The interactions between ants and these plants are more beneficial for the plants than for the ants. In the future, these examples need further investigations as they may be an initial stage of ant-plant interactions leading to the true mutualistic interactions.

Among several hypotheses about selective advantages of myrmecochory, there is one, which was not mentioned in the book because of a lack of information. This hypothesis may be formulated as «genetic diversity hypothesis». Since plant dispores are transported towards ant nests from different forest microsites, genetic variability of individuals of a myrmecochorous plant species should be the highest in areas where ants deposit waste diaspore bodies. Concentration of diaspores in the sites of higher genetic variability of individuals may prevent closely-related crossing beween individuals of the same plant species. This hypothesis still remains unproven.

Myrmecochory represents mutual relations between ants and plants. In the many-sided studies on this complex natural phenomenon, we tried to keep a balance between the plant perpective and the ant perspective. However, the ant's perspective turned out to be somewhat underrepresented. The main topic, which still has to be stressed in further field studies, is the measure of costs and advantages for various ant species in collecting and transporting diaspores of various myrmecochores.

Also non-myrmecochores have to be involved in these experiments, in order to find out how accidental relationships develop into potential mutual interrelationships. The question about the formation of mutual interactions between ants and plants still remains open. An initial advantage for the plant could be a general advantage, such as, for example, dispersal for distance and then a more specific advantage may appear. This advantage can lead to more specific interaction according to any aspect of the biology of both ants and plants. A detailed evolutionary analysis of dispore dispersal by ants within one plant group at the level of genus or family is neccessary to understand details of appearance of diaspore dispersal by ants in various plant groups.

Glossary

Anemochory *Diaspore dispersal* by birds (Pijl, 1982).

Autobarochory A type of *autochory*, when diaspores fall off through gravity (Levina, 1967).

Autochory *Diaspore dispersal*, which does not require dispersal agents (Levina, 1957).

Allochory *Diaspore dispersal*, in which dispersal agents, such as wind, water, animals, humans, are involved (Levina, 1957).

Aril A *seed* appendage consisting of sappy and pulpy tissues or presenting a kind of a pellicle or a fringe (Tahtadzhjan, 1985).

– **complex** An *aril* of a dual origin (Werker, 1997).

– **false** An *aril* originating in the region of *micropyle*, *chalaza* or along *raphe*-antiraphe (Tahtadzhjan, 1985).

– **micropylar** An *aril* originating in the region of *micropyle* (Tahtadzhjan, 1985).

– **raphal** An *aril* originating along *raphe*-antiraphe (Tahtadzhjan, 1985).

– **true** An *aril* originating on a *funicle* (Tahtadzhjan, 1985).

Arillode See *Aril false.*

Bract A somewhat modified leaf, which subtends a flower or *inflorescence* in its axis (Allaby, 1992a).

«Cafeteria» experiment A type of experiment, where food objects are presented to animals to record *removal rates* of the objects.

Capsule A polyspermous dehiscent fruit, usually dry, only in some cases sappy (Levina, 1967).

Carnivorous ants Ant species that feed mainly on animal food (Brian, 1983).

Caruncle See *Aril micropylar*.

Carpel A structure of the flower, in which *ovule*(s) is (are) developed, and collectively constitute the *gynoecium*.

Chalaza A basal region of the *ovule*, where integuments arise and the *funicle* is attached (Tahtadzhjan, 1985).

Chalazal process An *aril* originating in the region of *chalaza* (Tahtadzhjan, 1985).

Clone A plant or a group of plants derived from an ancestor by asexual reproduction (by detachment of roots, stems or leaves, or their parts) and developed into a complete plant.

Coenobium A *fruit* that disintegrates not only along the plane of *carpel* accretion, but also along additional septum perpendicular to the plane (Levina, 1967).

Colony A group of ant individuals, other than a single mated pair, which constructs nests or brings up offspring in a cooperative manner (Hölldobler and Wilson, 1990).

Cotyledon, cotyledonous leaf A *seed* leaf, which is borne on a plant embryo (Allaby, 1992a).

Diaspore A part of a plant that serves for reproduction and *dispersal* (Tahtadzhjan, 1985). Diaspores have different origins: *seeds, fruits, mericarps, erems,* etc.

Diaspore dispersal The departure of a diaspore from the *parent plant* (Howe and Smallwood, 1982).

Diaspore body A *diaspore* without *elaiosome*.

Diaspore predator An animal that eats or destroys a *diaspore* (Howe and Smallwood, 1988).

Diplochore A plant that uses a two-phase method of *diaspore dispersal*, e.g. *myrmecochory* follows *autochory* (Levina, 1957).

Dispersal of diaspores See *Diaspore dispersal.*

Dispersal syndrome A complex of plant adaptations to a certain *diaspore dispersal* method (Pijl, 1982).

Dissemination See *Diaspore dispersal.*

Dominant species The species having the most influence on community composition and form; also the most abundant species in the community (Allaby, 1994).

Dropping rate of diaspores The number of *diaspores* dropped by ants during transport along a certain distance of the route.

Elaiosome An appendage of a myrmecochorous *diaspore* that consists of parenchymatous cells containing a high amount of oils (Tahtadzhjan, 1985). Elaiosomes vary greatly in origin.

Endozoochory *Diaspore dispersal* through animals, internally (Pijl, 1982).

Erem A part of a disintegrating *fruit* of the *coenobium*-type (Levina, 1967).

Erem body An *erem* without an *elaiosome.*

Exocarp The outer layer of the *fruit* wall (Allaby, 1992a).

Epizoochory *Diaspore dispersal* through animals, externally (Pijl, 1982).

Flower tube A structure consisting of partly or fully accrete floral leaves.

Foot of the nest mound The wall around the *nest mound* composed of the soil removed from tunnels.

Forager A *worker* of an ant *colony* that goes in search of food (Hölldobler and Wilson, 1990).

Foraging ant See *Forager.*

Fruit A modified, owing to fertilisation, ripened *gynoecium* and its content including any other organs of a flower or *inflorescence*, with which it is inseparably connected (Allaby, 1992a).

Fruit body A fruit without elaiosome.

Funicle A stalk that attaches the *ovule* to the *placenta* (Tahtadzhjan, 1985).

Garbage dump A place of waste deposition of the ant *colony.*

Geocarpy The ripening of *diaspores* in soil (Levina, 1967).

Gynoecium Female reproductive organs of a flower comprising one or many *carpels* (Allaby, 1992).

Habitat A living place of an organism or community, characterised by its physical or biotic properties (Allaby, 1994).

Harvesting ants Ant species that feed mainly on *seeds* and store them in their nests (Hölldobler and Wilson, 1990).

Hilum A scar on a *seed* that marks the point, at which it was attached to the *funicle* or *placenta* (Tahtadzhjan, 1985).

Hydrochory *Diaspore dispersal* by water (Pijl, 1982).

Inflorescence A flowering structure that consists of more than a single flower (Allaby, 1992a).

Juvenile, juvenile plant A young immature plant with true leaves, either with *cotyledons* or without them.

Learning A feature of ant behaviour; the process of acquisition of knowledge about food source location by an individual ant worker.

Mandibles Paired structures of ant mouth-parts commonly used for seizing and cutting food (Allaby, 1992b).

Mature plant A fully grown and developed plant.

Mericarp A monospermous closed part of a disintegrating *fruit* that corresponds to a *carpel* (Levina, 1967).

Microhabitat A precise location within a *habitat*, where an individual species is found (Allaby, 1994).

Micropyle A canal in the integuments on the apical part of the ovule, through which the pollen tube (a tube formed from a germinating pollen grain) passes during fertilisation (Tahtadzhjan, 1985).

Microsite See *Microhabitat*.

Myrmecochore A plant, whose *diaspores* are dispersed by ants (Sernander, 1906).

– **facultative** a plant that has another method of *diaspore dispersal* in addition to myrmecochory (Ulbrich, 1928).

– **obligate** a plant that uses only one method of *diaspore dispersal* – myrmecochory (Ulbrich, 1928).

Myrmecochorous plant See *Myrmecochore.*

Myrmecochorous syndrome A complex of special morphological, anatomical, biochemical, and phenological adaptations of plants that enhance ant attraction and thus increase the effectiveness of *myrmecochory.*

Myrmecochory One of *diaspore dispersal* methods; *diaspore dispersal* by ants (Ulbrich, 1928).

– non-specialised A type of *myrmecochory*, where *diaspores* lack fat-containing appendages.

– specialised A type of *myrmecochory*, where *diaspores* have special fat-containing bodies – *elaiosomes.* Ants consume *elaiosomes* and disperse *diaspores.*

Myrmecophily Mutually beneficial interactions between ants and plants (Beattie, 1983).

Mutualism, mutualistic relations Interactions between members of two species, which benefit both (Allaby, 1994).

Nest mound A part of an ant nest that is constructed of a mound of soil and projects above the ground surface.

Niche The combined range of environmental conditions, such as temperature, humidity, and food resources, within which a species can exist and reproduce (Hölldobler and Wilson, 1990).

Ornithochory *Diaspore dispersal* by birds (Pijl, 1982).

Ovule A structure in seed plants that, after fertilisation, develops into a *seed* (Allaby, 1992a).

Parent plant A plant, which is producing or has produced *diaspores.*

Placenta A part of the ovary (a basal part of *gynoecium*) wall carrying the *ovules* (Tahtadzhjan, 1985).

Pericarp A *fruit* wall, often with three distinct layers (Tahtadzhjan, 1985).

Perigonium Floral leaves that are not differentiated into sepals and petals.

Pseudoaril See *Aril false.*

Raphe A part of the *funicle* of an anatropous *ovule* adnate to the integuments and forming a ridge from the *hilum* to *chalaza* (Tahtadzhjan, 1985).

Receptacle A part of the stem, from which all the parts of the flower arise (Allaby, 1992a).

Recruitment A special form of assemblage, in which members of an ant *colony* are directed to some point in space, where work is required (Hölldobler and Wilson, 1990).

Redispersal of diaspores The transference of myrmecochorous *diaspores* by ants from nests after consumption of *elaiosomes*.

Redistribution of diaspores See *Redispersal of diaspores.*

Relocation of diaspores See *Redispersal of diaspores.*

Removal rate of diaspores The number of *diaspores* removed by ants over a certain period of time.

Sarcotesta The outer and usually soft and sappy part of the *seed* cover (Tahtadzhjan, 1985).

Secondary relocation of diaspores See *Redispersal of diaspores.*

Seed A dispersal unit of seed plants. It is formed on the *parent plant* from a fertilised and ripened *ovule* and contains an embryo and a food store enclosed in a protective outer coat (Tahtadzhjan, 1985).

Seed body A *seed* without *elaiosome.*

Seed flow *Diaspore* movement from the *parent plant* to the germination site. It can be caused either by the plant itself or by dispersal agents, such as wind, water, animals, and humans.

Seed producing capacity The number of *seeds* produced by a plant during one season.

Seedling A young immature plant grown and developed from a *seed*, with *cotyledons*.

Selective advantages of myrmecochory Advantages, which are promoted by natural selection and used by plants dispersed by ants.

Soil seed bank Ungerminated but viable *seeds* that lie in the soil (Allaby, 1994).

Spermoderm A protective outer coat of the *seed* (Tahtadzhjan, 1985).

Strophiole See *Aril raphal*.

Style An extension of the *carpel* that connects the *ovary* to the stigma (a part of the female reproductive organs on which pollen grains germinate) (Allaby, 1992a).

Time of half removal of diaspores The time, in which half of the *diaspores* are removed from the complete diaspore depot.

Worker, working individual A member of the non-reproductive labouring caste of the ant *colony* that are engaged in brood care, forage, building of nests etc. (Hölldobler and Wilson, 1990).

Zoochory *Diaspore dispersal* by animals (Levina, 1957).

References

Abbott I., Heurck P. Comparison of insects and vertebrates as removers of seed and fruit in a Western Australian forest // Austral. J. Ecol. – 1985. – V. 10. – P. 165 – 168.

Adlerz G. Myrmecologiska studier. 2. Svenska myror och deras lefnadsfoerhallanden. – Stockholm: K.V.A. Bihang, 1886. – Bd. 2, № 18 (cited after Sernander, 1906).

Affre L., Thompson J.D., Debussche M. The reproductive biology of the Mediterranean endemic *Cyclamen balearicum* Willk. (Primulaceae) // Bot. J. Linn. Soc. – 1995. – V. 118. – P. 309 – 330.

Allaby M. (ed.) The concise Oxford dictionary of botany. – Oxford, New York: Oxford Univ. Press, 1992a. – 442 p.

Allaby M. (ed.) The concise Oxford dictionary of zoology. – Oxford, New York: Oxford Univ. Press, 1992b. – 508 p.

Allaby M. (ed.) The concise Oxford dictionary of ecology. – Oxford, New York: Oxford Univ. Press, 1994. – 415 p.

Andersen A.N. Seed removal by ants in the mallee of North Western Victoria // Ant-plant interactions in Australia / Ed. R.C. Buckley. – The Hague: Dr. W. Junk, 1982. – P. 31 – 43.

Andersen A.N. Dispersal distance as a benefit of myrmecochory // Oecologia. – 1988. – V. 75. – P. 507 – 511.

Andre E. Species des hymenopteres d'Europe et d'Algerie. T. 2. Formicides, Vespides. – Paris, 1886 (cited after Sernander, 1906).

Artjushenko Z.T. Amaryllidaceae Jaume St.-Hilaire of the USSR: morphology, sytematics and utilization. – Leningrad: Nauka, 1970. – 180 p. (in Russian).

Ashton D.H. Seed harvesting by ants in forests of *Eucalyptus regnans* F. Muell. in Central Victoria // Austral. J. Ecol. – 1979. – V. 4, № 3. – P. 265 – 277.

Auld T.D. Population dynamics of the shrub *Acacia suaveolens* (Sm.) Willd.: Dispersal and the dynamics of the soil seed-bank // Austral. J. Ecol. – 1986. – V. 11 – P. 235 – 254.

Baroni-Urbani C., Nielsen M.G. Energetics and foraging behaviour of the European seed harvesting ant *Messor capitatus*. II. Do ants optimise their harvesting? // Physiol. Entomol. – 1990. – V. 15. – P. 449 – 461.

Baciu V.A. Cercetari asupra originei si desvoltarii elaiosomului la Boraginee I. // Bul. Fac. de Stiinte din Cernauti, Univ. Romanea. – 1931. – Bd. 5. – S. 48 – 63.

Baciu V.A. Neue Untersuchungen über die Ablösung der Achänen von *Centaurea* Linne. // Bul. Fac. de Stiinte din Cernauti, Univ. Romanea. – 1932. – Bd. 6. – S. 138 – 141.

Baciu V.A. Anatomia fructelor de Boraginee, cu accentuarea ontogenezei elaiosomului si a detasarii mericarpiilor // Bul. Fac. de Stiinte din Cernauti, Univ. Romanea. – 1935. – Bd. 9. – S. 123 – 169.

Beattie A.J. Distribution of ant-dispersed plants // Sonderbl. Naturwiss. Ver. Hamburg. – 1983. – V. 7. – P. 249 – 270.

Beattie A.J. The evolutionary ecology of ant-plant mutualisms. – Cambrodge, London, New Jork etc.: Cambridge Univ. Press, 1985. – 182 p.

Beattie A.J., Culver D.C. The nest chemistry of two seed-dispersing ant species // Amer. Midland Nat. – 1977. – V. 97. – P. 390 – 399.

Beattie A.J., Culver D.C. Neighbourhood size in *Viola* // Evolution. – 1979. – V. 33, № 4. – P. 1226 – 1229.

Beattie A.J., Culver D.C. The guild of myrmecochores in the herbaceous flora of West Virginia forests // Ecology. – 1981. – V. 62, № 1. – P. 107 – 115.

Beattie A.J., Culver D.C. Inhumation: how ants and other invertebrates help seeds // Nature. – 1982. – V. 297. – P. 627.

Beattie A.J., Culver D.C. The nest chemistry of two seed-dispersing ant species // Oecologia. – 1983. – V. 56. – P. 99 – 103.

Beattie A., Lyons N. Seed dispersal in *Viola*: Adaptations and strategies // Amer. J. Bot. – 1975. – V. 62 – P. 714 – 722.

Berg R.Y. Development and dispersal of the seed of *Pedicularis silvatica* // Nytt. Mag. Bot. – 1954. – V. 2. – P. 1 – 59.

Berg R.Y. Seed dispersal, mophology and phylogeny of *Trillium* // Skr. norske Vid. Acad. Oslo, Math. Nat. Kl. – 1958. – № 1. – P. 1 – 36.

Berg R.Y. Seed dispersal of *Dendromecon*: its ecologic, evolutionary, and taxonomic significance // Amer. J. Bot. – 1966. – V. 53, № 1. – P. 61 – 73.

Berg R.Y. Adaptation and evolution in *Dicentra* (Fumariaceae), with special reference to seed, fruit, and dispersal mechanism // Nytt. Mag. Bot. – 1969. – V. 16. – P. 49 – 75.

Berg R.Y. Dispersal ecology of *Vancouveria* (Berberidaceae) // Amer. J. Bot. –1972. – V. 59. – P. 109 – 122.

Berg R.Y. Myrmecochorous plants in Australia and their dispersal by ants // Austral. J. Bot. – 1975. – V. 23. – P. 475 – 508.

Bernard F. Comportement de la fourmi *Messor barbara* (L.) pour la recolte des graines de *Trifolium stellatum* L. // Bull. Soc. Hist. Natur. Afr. Nord. – 1971. – V. 62, №№ 1 – 2. – P. 15 – 19.

Bisseau de J.C., Pasteels J.M. Regulated food recruitment through individual behavior of scouts in the ant, *Myrmica sabuleti* (Hymenoptera: Formicidae) // J. Insect Behav. – 1994. – V. 7, № 6. – P. 767 – 777.

Bjoernstad I.N. A revision of the genus *Pancratium* L. (Amaryllidaceae) in Africa South of the Sahara // Norw. J. Bot. – 1973. – V. 20, № 4. – P. 281 – 291.

Blatter E. Myrmecosymbiosis in the Indo-Malayan Flora // J. Indian Bot. Soc. – 1928. – V. 7. – P. 176 – 185.

Bond W.J., Stock W.D. The costs of leaving home: ants disperse myrmecochorous seeds to low nutrient sites // Oecologia. – 1989. – V. 81. – P. 412 – 417.

Bond W.J., Yeaton R., Stock W.D. Myrmecochory in Cape fynbos // Ant-plant interactions / Eds Huxley C.R., Cutler D.F. – Oxford, N. Y., Tokyo: Oxford Univ. Press, 1991. – P. 448 – 462.

Bor J., Kapil R.N. *Euphorbia geniculata*, ovule to seed // Acta Bot. Neer. – 1976. – V. 24, №№ 3 – 4 (cited after Tahtadzhjan, 1992).

Bregman R. Forms of seed dispersal in Cactaceae // Acta Bot. Neerl. – 1988. – V. 37, № 3. – P. 395 – 402.

Bresinsky A. Bau, Entwicklungsgeschichte und Inhaltsstoffe der Elaiosomen // Bibl. Bot. – 1963. – Bd. 126. – S. 1 – 54.

Brew C.R., O'Dowd D.J., Rae I.D. Seed dispersal by ants: behaviour-releasing compounds in elaiosomes // Oecologia. – 1989. – V. 80. – P. 490 – 497.

Brian M.V. Social Insects. Ecology and behavioural biology. – London, New-York: Chapman and Hall, 1983.

Brouwer W., Staehlin A. Handbuch des Samenkunde für Landwirtschaft, Gartenbau und Forstwirtschaft. – Frankfurt-am-Main, 1955 (cited after Tahtadzhjan, 1992).

Buckley S.B. *Myrmica (Atta) molefaciens*, «stinging ant», or «mound-making ant», of Texas. // Proceed. Acad. Nat. Sci., Philadelphia. – 1860. – P. 445 – 447.

Bullock S.H. Seed dispersal of *Dendromecon* by the seed predator *Pogonomyrmex* // Madrono. – 1974. – V. 22, № 7 (cited after Tahtadzhjan, 1988).

Bullock S.H. Life history and seed dispersal of the short-lived chaparral shrub *Dendromecon rigida* (Papaveraceae) // Amer. J. Bot. – 1989. – V. 76, № 10. – P. 1506 – 1517.

Buxbaum F. Morphology of cacti. Section 3. Fruits and seeds. – Pasadena, Calif.: Abbey Garden Press, 1955 (cited after Roth, 1977).

Byrne M.M., Levey D.J. Removal of seeds from frugivore defecations by ants in a Costa Rican rain forest // Vegetatio. – 1993. – V. 108. – P. 363 – 374.

Cammaerts M.-C. Recruitment to food in *Myrmica rubra* L. // Biol. Behav. – 1978. – V. 4, № 2. – P. 159 – 172.

Carroll C.R., Risch S.J. The dynamics of seed harvesting in early successional communities by a tropical ant, *Solenopsis geminata* // Oecologia. –1984. – V. 61. – P. 388 – 392.

Carthy J.D. The orientation of two allied species of British ant. I. Visual direction finding in *Acanthomyops (Lasius) niger* // Behaviour. – 1951. – V. 3. – P. 275 – 303.

Cerdan P., Delye G. La consommation des graines par les fourmis *Messor barbarus* et *Messor sanctus* dans la plaine de Crau (Bouches du Rhone) // Actes Colloq. Insectes Soc. – 1990. – V. 6. – P. 165 – 171.

Chadab R., Rettenmeyer C.W. Mass recruitment by army ants // Science. – 1975. – V. 188. – P. 1124 – 1125.

Clay K. Myrmecochory in the tailing arbutus (*Epigaea repens* L.) // Bull. Torrey Bot. Club. – 1983. – V. 110. – P. 166 – 169.

Corff J.L., Horvitz C.C. Dispersal of seeds from chasmogamous and cleistogamous flowers in an ant-dispersed neotropical herb // Oikos. – 1995. – V. 73. – P. 59 – 64.

Corner E.J.H. The seeds of Dicotyledons. – Cambridge: Cambridge Univ. Press, 1976. – V. 2. (cited after Werker, 1997).

Crété P. A propos de la strophiole du *Chelidonium majus* L. // Bull. Soc. Bot. France. – 1944. – V. 91, №№ 1 – 2 (cited after Tahtadzhjan, 1988).

Culver D., Beattie A. Myrmecochory in *Viola*: Dynamics of seed-ant interactions in some West Virginia species // J. Ecol. – 1978. – V. 66. – P. 53 – 72

Culver D., Beattie A. The fate of *Viola* seeds dispersed by ants // Amer. J. Bot. – 1980. – V. 67. – P. 710 – 714.

Culver D.C., Beattie A.J. Effects of ant mounds on soil chemistry and vegetation patterns in a Colorado Montane meadow // Ecology. – 1983. – V. 64, № 3. – P. 485 – 492.

Cushman J. H., Beattie A. Mutualisms: assesing the benefits to hosts and visitors // Tree. – 1991. – V. 6. – P. 193 – 195.

Czervinski A., Jakubczyk H., Petal J. Influence of ant hills on the meadow soil // Pedobiologia – 1971. – V. 11. – P. 277 – 285.

Dadd R.H. Insect nutrition: Current developments and metabolic implications // Ann. Rev. Entomol. – 1973. – V. 18. – P. 381 – 429.

Danin A., Yom-Tov Y. Ant nests as primary habitats of *Silybum marianum* (Compositae) // Pl. Syst. Evol. – 1990. – V. 169. – P. 209 – 217.

Davidson D. W. Ecological studies of Neotropical ant gardens // Ecology. – 1988. – V. 69. – P. 1138 – 1152.

Davidson D.W., Fisher B.L. Symbiosis of ants with *Cecropia* as a function of light regime // Ant-plant interactions / Eds. Huxley C.R., Cutler D.F. – Oxford, New York, Tokyo: Oxford Univ. Press, 1991. – P. 289 – 309.

Davidson, D.W., Foster R.B., Snelling R.R., Lozada P.W. Variable composition of some tropical ant-plant symbioses // Plant-animal interactions: evolutionary ecology in tropical and temperate regions. – New York, Chichester etc.: Wiley, 1991. – P. 145 – 162.

Davidson D.W., Morton S.R. Competition for dispersal in ant-dispersed plants // Science. – 1981a. – V. 213. – P. 1259 – 1261.

Davidson D.W., Morton S.R. Myrmecochory in some plants (F. Chenopodiaceae) of the Australian arid zone // Oecologia. – 1981b. – V. 50. – P. 357 – 366.

Davidson D.W., Morton S.R. Dispersal adaptations of some *Acacia* species in the Australian arid zone // Ecology. – 1984. – V. 65, № 4. – P. 1038 – 1051.

Davidson D.W., Seidel J.L., Epstein W.W. Neotropical ant gardens 2. Bioassays of seed compounds // J. Chem. Ecol. – 1990. – V. 16. – P. 2993 – 3013.

Dean W.R.J., Yelton R.I. The importance of harvester ant *Messor capensis* nest-mounds as germination sites in the southern Karoo, South Africa // Afr. J. Ecol. – 1992. – V. 30. – P. 335 – 345.

Delpino F. Sul fenomeno della macrobiocarpia in alcune piante // Rendic. R. Acad. Sc. Fis. e Matem., Napoli, Ser. 3. – 1903. – V. 9. – P. 48 – 57.

Dethier V.G. Chemical insect attractants and repellents. – Blakiston, Philadelphia, 1947 (cited after Marshall et al., 1979).

Dlussky G.M. Ants of the genus *Formica* (Hymenoptera, Formicidae, G. *Formica*). – Moskwa: Nauka, 1967. – 236 p. (in Russian).

Dobrzanska J. The control of the territory by *Lasius fuliginosus* Latr. // Acta Biologiae Experimentalis. – 1966. – V. 26, № 2. – P. 193 – 213.

Drake W.E. Ant-seed interaction in dry sclerophyll forest on north Stradbroke Island, Queensland // Austral. J. Bot. – 1981. – V. 29. – P. 293 – 309.

Espadaler X., Gomes C. Seed production, predation and dispersal in the Mediterranean myrmecochor *Euphorbia characias* (Euphorbiaceae) // Ecogeography. – 1996. – V. 19. – P. 7 – 15.

Federle W., Fiala B., Maschwitz U. *Camponotus (Colobopsis)* (Mayr 1861) and *Macaranga* (Thouars 1806): a specific two-partner ant-plant system from Malaysia // Tropical Zoology. – 1998. – V. 11. – P. 83 – 94.

Forel A. Les Fourmis de la Suisse // Nouv. Mem. Soc. Helv. Sci. Nat. Zuerich. – 1874. – V. 26. – P. 1 – 447.

Fourcassie V., Traniello J.F.A. Effects of experience on food-searching behavior in the ant *Formica schaufussi* (Hymenoptera: Formicidae) // J. Insect Behav. – 1993. – V. 6, № 3. – P. 287 – 299.

Ganeshaiah K., Shaanker R.U. Evolution of a unique seed maturity pattern in *Croton bonplandianum* Baill strengthens ant-plant mutualism for seed dispersal // Oecologia. – 1988. – V. 77. – P. 130 – 134.

Gates B.N. The dissemination by ants of the seeds of blood root, *Sanguinaria canadensis* // Rhodora. – 1942. – V. 44 (cited after Tahtadzhjan, 1988).

Gates B.N. Carunculate seed dissemination by ants // Rhodora. – 1943. – V. 45 (cited after Tahtadzhjan, 1988).

Geitler L. Der Bau der Riesenkerne des Elaiosoms von *Corydalis cava* // Chromosoma. – 1944. – Bd. 2. – S. 544 – 548.

Geitler L. Ergebnisse und Probleme der Endomitoseforschung // Oesterr. Bot. Z. – 1949. – Bd. 95. – S. 277 – 299.

Geitler L. Notizen zur endomitotischen Polyploidisierung in Trichocyten und Elaiosomen sowie über Kernstrukturen bei *Gagea lutea* // Chromosoma. – 1950. – Bd. 3. – S. 271 – 281.

Gibson W. Selective advantages to hemi-parasitic annuals, genus *Melampyrum*, of a seed-dispersal mutualism involving ants: 1. Favorable nest sites // Oikos. – 1993a. – V. 67. – P. 334 – 344.

Gibson W. Selective advantages to hemi-parasitic annuals, genus *Melampyrum*, of a seed-dispersal mutualism involving ants: 2. Seed-predator avoidance // Oikos. – 1993b. – V. 67.– P. 345 – 350.

Gilbert L.I. Lipid metabolism and function in insects // Adv. Insect. Physiol. – 1967. – V. 4. – P. 69 – 211.

Gilbert L.I., Chino H. Transport of lipids in insects // J. Lipid. Res. – 1974. – V. 15. – P. 439 – 456.

Greenfeld E.K. Influence of ants on soil reaction // Zool. Zh. – 1941. – V. 20, № 1. – P. 100 (in Russian).

Grimalsky V.I. About the role of red forest ants (*Formica rufa*) in forest biocoenisis in Left-bank Marshy Woodlands of Ukraine // Zool. Zh. – 1960. – V. 39. – P. 394 – 398 (in Russian).

Gorb O.V. Seed morphology and seed dispersal in two *Corydalis* species // Ukr. Bot. Zh. – 1998. – V. 55, № 1. – P. 62 – 66.

Gorb O.V. Effect of deposition of seeds of myrmecochorous plants on seed dispersal by ants // Ukr. Bot. Zh. – 2000. – V. 57, № 3. – P. 332 – 336 (in Ukrainian).

Gorb E.V., Gorb S.N. Competition between two *Corydalis* species for ants, which disperse their seeds // Abstr. Thes. 9th Int. Symposium on Insect-Plant Relationships, June 24 – 30, Gwatt, Switzerland. – 1995a. – P. 86.

Gorb S.N., Gorb E.V. Removal rates of seeds of five myrmecochorous plants by ants *Formica polyctena* // Oikos. – 1995b. – V. 73. – P. 367 – 374.

Gorb E.V., Gorb S.N. Influence of myrmecochory on spatial pattern and seedlings mortality in *Asarum europaeum* L. (Aristolochiaceae) // Zh. Obshch. Biol. – 1995c. – V. 56, № 4. – P. 470 – 476 (in Russian).

Gorb E.V., Gorb S.N. Ant nests as habitats of myrmecochores: do the nest mounds of *Formica polyctena* Fabr. provide the advantageous microsites for the violet *Viola odorata* L.? // Zh. Obshch. Biol. – 1996. – V. 57, № 4. – P. 452 – 456 (in Russian).

Gorb E.V., Gorb S.N. The decrease in seedling density as a selective advantage of myrmecochory in *Corydalis solida* (L.) Clairv. (Fumariaceae) // Zh. Obshch. Biol. – 1997. – V. 58, № 5. – P. 89 – 96 (in Russian).

Gorb E.V., Gorb S.N. Myrmecochory in broadleaved forest: the effect of ants on seed withdrawal // Abstr. Thes. 9th Russ. Myrmecolog. Symposium «Ants and forest defence», 24 – 28 August 1998, Peshki. – Moscow, 1998a. – P. 91 – 94 (in Russian).

Gorb S.N., Gorb E.V. Seed bank of the ant mounds as the result of the seed flow on the territories *Formica polyctena* Fabr. in deciduous forest // Ecologia. – 1998b. – № 1. – P. 62 – 70 (in Russian).

Gorb S.N., Gorb E.V. Dropping rates of elaiosome-bearing seeds during transport by ants (*Formica polyctena* Foerst.): Implications for distance dispersal // Acta Oecologica.– 1999a.– V. 20, № 5. – P. 509 – 518.

Gorb S.N., Gorb E.V. Effects of ant species composition on seed removal in deciduous forest in eastern Europe // Oikos.– 1999b.– V. 84.– P. 100 – 118.

Gorb E., Gorb S. Effects of seed aggregation on the removal rates of elaiosome-bearing *Chelidonium majus* and *Viola odorata* seeds carried by *Formica polyctena* ants // Ecological Research. – 2000. – V. 15, № 2. – P. 187 – 192.

Gorb E.V., Gorb S.N. Effect of some characters of myrmecochorous syndrome on the removal rates of diaspores of *Corydalis cava* and *Pulmonaria obscura* by *Formica polyctena* ants // Zh. Obshch. Biol. – 2001a. – V. 62, № 2. – P. 132 – 140 (in Russian).

Gorb E.V., Gorb S.N. Myrmecochorous syndrome // Zh. Obshch. Biol. – 2001b. – V. 62, № 6. – P. 496 – 511 (in Russian).

Gorb E., Gorb S. Contact separation force of the fruit burrs in four plant species adapted to dispersal by mechanical interlocking // Plant Physiology and Biochemistry. – 2002. – V. 40, № 4. – P. 373 – 381.

Gorb S.N., Gorb E.V., Puntilla P. Effects of redispersal of seeds by ants on the vegetation pattern in a deciduous forest: A case study // Acta Oecologica. – 2000. – V. 21, №№ 4 – 5. – P. 293 – 301.

Gorb S., Gorb E., Sindarovskaya Yu. Interaction between the non-myrmecochorous herb *Galium aparine* and the ant *Formica polyctena* // Plant Ecology. – 1997. – V. 131. – P. 215 – 221.

Gorb E.V., Panasenko N.V., Gorb S.N. Adaptations of the *Pulmonaria obscura* Dumort. (Boraginaceae) fruits to dispersal by ants // Vestn. Ecol. – 1996. – №№ 1 – 2. – P. 58 – 61 (in Russian).

Griffith C.G. Distribution of *Viola blanda* in relation to within-habitat variation in canopy openness, soil phosphorus, and magnesium // Bull. Torr. Bot. Club. – 1996. – V. 123, № 4. – P. 281 – 285.

Grime J. P. Evidence for the existence of three primary strategies in plants and its revelance to ecological and evolutionary theory // Amer. Nat.– 1977.– V. 111. – P. 1169 – 1194.

Gunther R.W., Lanza J. Variation in attractiveness of *Trillium* diaspores to a seed dispersing ant // Amer. Midland Nat. – 1989. – V. 122. – P. 321 – 328.

Handel S.N. Dispersal ecology of *Carex pedunculata* (Cyperaceae), a new north American myrmecochore // Amer. J. Bot. – 1976. – V. 63. – P. 1071 – 1079.

Handel S.N. The competitive relationship of three woodland sedges and its bearing on the evolution of ant-dispersal of *Carex pedunculata* // Evolution. – 1978a. – V. 32. – P. 151 – 163.

Handel S.N. New ant-dispersed species in genera *Carex, Luzula* and *Claytonia* // Canad. J. Bot. – 1978b. – V. 56. – P. 2925 – 2927.

Handel S.N., Beattie A.J. Seed dispersal by ants // Sci. Amer. – 1990. – № 8. – P. 58 – 64.

Handel S.N., Fisch S.B., Schatz G.E. Ants disperse a majority of herbs in a mesic forest community in New York State // Bull. Torr. Bot. Club. – 1981. – V. 108, № 4. – P. 403 – 437.

Hansen S.R. Resource utilization and coexistence of three species of *Pogonomyrmex* ants in an Upper Sonoran grassland community // Oecologia. – 1978. – V. 35. – P. 109 – 117.

Heithaus E.R. Seed predation by rodents on three ant-dispersed plants // Ecology. – 1981. – V. 62, № 1. – P. 136 – 145.

Heithaus E.R. Seed dispersal mutualism and the population density *of Asarum canadense*, an ant-dispersed plant // Frugivores and seed dispersal / Eds. Estrada A., Fleming T.H. – Dordrecht: Dr. W. Junk Publishers, 1986. – P. 199 – 210.

Heithaus E.R., Culver D.C., Beattie A.J. Models of some ant-plant mutualisms // Amer. Nat. – 1980. – V. 116. – P. 347 – 361.

Higashi S., Tsuyuzaki S., Ohara M., Ito F. Adaptive advantages of ant-dispersed seeds in the myrmecochorous plant *Trillium tschonoskii* (Liliaceae) // Oikos. – 1989. – V. 54. – P. 383 – 394.

Higashi S., Ito F. Ground beetles and seed dispersal of the myrmecochorous plant *Trillium tschonoskii* (Trilliaceae) // Ant-plant interactions / Eds. Huxley C.R., Cutler D.F. – Oxford, New York, Tokyo: Oxford University Press, 1991. – P. 487 – 492.

Hocking P.J., Kortt A.A. Growth and nutrient acumulation by fruits of the perennial legume, *Hardenbergia violacea*, with special reference to myrmecochory // New Phytol. – 1987. – V. 105. – P. 89 – 102.

Hölldobler B., Wilson E.O. The ants. -Cambridge: The Belknap Press of Harvard Univ. Press, 1990. – 732 p.

Holt S.J. On the foraging activity of wood ant // J. Anim. Ecol. – 1995. – V. 24. – P. 1 – 34.

Holttum R.E. The Marantaceae of Malaya // Gard. Bull. Singapore. – 1951. – V. 13. – P. 263 – 275.

Horstmann K. Freilanduntersuchungen zum Rekrutierungsverhalten bei der Waldameise *Formica polyctena* Foerster. // Waldhygiene. – 1975. – Bd. 11. – S. 33 – 40.

Horvitz C.C. Light environments, stage structure, and dispersal syndromes of Costa Rican Marantaceae // Ant-plant interactions / Eds. Huxley C.R., Cutler D.F. – Oxford, N. Y., Tokyo: Oxford Univ. Press, 1991. – P. 463 – 485.

Horvitz C.C., Schemske D.W. Ant-nest soil and seedling growth in a neotropical ant-dispersed herb // Oecologia. – 1986a. – V. 70. – P. 318 – 320.

Horvitz C.C., Schemske D.W. Seed dispersal of a neotropical myrmecochore: Variation in removal rates and dispersal distance // Biotropica. – 1986b. – V. 18, № 4. – P. 319 – 323.

Howe H.F., Smallwood J. Ecology of seed dispersal // Ann. Rev. Ecol. Syst. – 1982. – V. 13. – P. 201 – 228.

Howe H.F., Westley L.C. Ecological relationships of plants and animals. – New York, Oxford: Oxford Univ. Press, 1988.

Huber H. Die Samenmerkmale und Verwandtschaftsverhältnisse der Liliifloren // Mitt. Bot. Staatssamml. München. – 1969. – Bd. 8, № 2. – S. 219 – 538.

Hughes L. The relocation of ant nest entrances: Potential consequences for ant-dispersed seeds // Austral. J. Ecol. – 1991. – V. 16.– P. 207 – 214.

Hughes L., Westoby M. Removal rates of seeds adapted for dispersal by ants // Ecology. – 1990. – V. 71. – P. 138 – 148.

Hughes L., Westoby M. Effect of diaspore characteristics on removal of seeds adapted for dispersal by ants // Ecology. – 1992a. – V. 73, № 4. – P. 1300 – 1312.

Hughes L., Westoby M. Fate of seeds adapted for dispersal by ants in Australian sclerophyl vegetation // Ecology. – 1992b. – V. 73, № 4. – P. 1285 – 1299.

Hughes L., Westoby M., Jurado E. Convergence of elaiosomes and insect prey: evidence from ant foraging behaviour and fatty acid composition // Funct. Ecol. – 1994. – V. 8. – P. 358 – 365.

Huxley C.R. The ant-plants *Myrmecodia* and *Hydnophytum* (Rubiaceae), and the relationships between their morphology, ant occupants, physiology and ecology // New Phytologist. – 1978. – V. 80. – P. 231 – 268.

Imhanitskaja N.N. Violaceae // In: Plants' life / Ed. Tahtadzhjan A.L. – Moskwa: Prosveshchenije, 1981. – V. 5(2). – P. 40 – 45 (in Russian).

Jansen D.H. Epiphytic myrmecophytes in Sarawaki: Mutualism through the feeding of plants by ants // Biotropica. – 1974. – V. 6. – P. 237 – 259.

Johnson R.A. Learning, memory, and foraging efficiency in two species of desert seed-harvester ants // Ecology. – 1991. – V. 72. – P. 1408 – 1419.

Kapil R.N. A further contribution to the morphology and life history of *Chrozophora* Neck. // Phytomorphology. – 1956a. – V. 6, №№ 3 – 4. – P. 278 – 288.

Kapil R.N. Development of embryosac and endosperm in *Chrozophora rottleri*: a reinvestigation // Bot. Gaz. – 1956b. – V. 117 (cited after Tahtadzhjan, 1988).

Kapil R.N., Bor J., Bouman F. Seed appendages in angiosperms. I. Introduction // Bot. Jahrb. Syst. – 1980. – Bd. 101. – S. 555 – 573.

Kaspari M Removal of seeds from neotropical frugivore droppings: Ant responses to seed number // Oecologia. – 1993. – V. 95.– P. 81 – 88.

Kennedy H. Systematics and pollination of the «closed flowered» species of Calathea (Marantaceae) // Univ. Calif. Berkley Publs Bot. – 1978. – V. 71. – P. 14 – 79.

Kerner A. Pflanzenleben: In 2 Bd. – Aufl. Lpz. und Wien, 1896. – Bd. 1 (cited after Sernander, 1906).

Kerner A. Pflanzenleben: In 2 Bd. – Aufl. Lpz. und Wien, 1898. – Bd. 2 (cited after Sernander, 1906).

King T.J. The plant ecology of ant-hills in calcareous grasslands. I. Patterns of species in relation to ant-hills in southern England // J. Ecol. – 1977a. – V. 65. – P. 235 – 256.

King T.J. The plant ecology of ant-hills in calcareous grasslands. II. Succession on the mounds // J. Ecol. – 1977b. – V. 65. – P. 257 – 278.

King T.J. The plant ecology of ant-hills in calcareous grasslands. III. Factors affecting the population sizes of selected species // J. Ecol. – 1977c. – V. 65. – P. 279 – 315.

Kirchner W. Jahreszyklische Untersuchungen zur Reservestoffspeicherung und Überlebensfähigkeit adulter Waldameisenarbeiterinnen (gen. *Formica*, Hym., Formicidae) // Zool. Jahrb. Abt. I. – 1964. – V. 71, № 1. – P. 1 – 72.

Kjellsson G. Seed fate in a population of *Carex pilulifera* L. I. Seed dispersal and ant seed mutualism // Oecologia. – 1985. – V. 67. – P. 416 – 423.

Kleinfeldt S. E. Ant-gardens: mutual exploitation // Insects and the plant surface. – London, Sydney, Baltimore: Edward Arnold Ltd., 1986. – P. 283 – 294.

Komar G.A. Arils, their morhhological nature, texture and function // Bot. Zh. – 1965. – V. 50, № 5. – P. 715 – 724 (in Russian).

Komar G.A. Arilloids in some *Scilla* species// Bot. Zh. – 1973. – V. 58, № 10. – P. 1503 – 1508 (in Russian).

Komar G.A. Arils in representatives of genus *Chionodoxa* // Bot. Zh. – 1974.– V. 59, № 11. – P. 1617 – 1622 (in Russian).

Komar G.A. Arils and aril-like structures in some representatives of Liliales // Bot. Zh. – 1978.– V. 63, № 7. – P. 935 – 955 (in Russian).

Kostrin K.A. Influence of ants on growth of brier and raspberry // Priroda. – 1951. – № 2. – P. 94 (in Russian).

Kuntze O. Die Schutzmittel der Pflanzen gegen Tiere und Wetterungunst // Lpz., Gratisbeilage zur Botan. Zeitung. – 1877 (cited after Sernander, 1906).

Lagerheim G. Über *Lasius fuliginosus* (Latr.) und seine Pilzzucht // Entomol. Tidskr., Stockholm. – 1900. – Agr. 21(cited after Sernander, 1906).

Landes M. Seed development in *Acalypha rhomboidea* and some other Euphorbiaceae // Amer. J. Bot. – 1946. – V. 33, №№ 6 – 7. – P. 562 – 568.

Lange A.H. Effect of the sarcotesta on germination of *Carica papaya* // Bot. Gaz. – 1961. – V. 122. – P. 305 – 311.

Lebedev A.A. Original symbiosis // Priroda. – 1949. – № 4. – P. 74 – 75 (in Russian).

Letourneau D. Ants, stem-borers, and fungal pathogens: Experimental tests of a fitness advantage in *Piper* ant-plants // Ecology. – 1998. – V. 79. – P. 593 – 603.

Levey D.J., Byrne M.M. Complex ant-plant interactions: rain forest ants as secondary dispersers and post-dispersal seed predators // Ecology. – 1993. – V. 74. – P. 1802 – 1812.

Levina R.E. Methods of dispersal of seeds and fruit. – Moskwa: Izdatelstvo Moskowskogo Gosudarstvennogo Universiteta, 1957. – 360 p. (in Russian).

Levina R.E. Fruits: morphology, ecology, practical significance. – Saratov: Privolzhskoje Knizhnoje Izdatelstvo, 1967. – 215 p. (in Russian).

Lincecum G. On the agricultural ant of Texas (*Myrmica molefaciens*) // Proceed. Acad. Nat. Sci., Philadelphia. – 1866. – P. 323 – 331.

Logoa A.M.M.A., Pereira M.F.A. The role of the caruncle in the germination of the seeds of *Ricinus communis* // Plant Physiol. Biochem. – 1987. – V. 25. – P. 125 – 128.

Lubbock J. Ants, bees, and wasps.– London, 1882 (cited after Sernander, 1906).

Ludwig F. Biologische Beobachtungen an *Helleborus foetidus* // Österr. Bot. Zeitschr., Wien. – 1898. – Bd. 48. – S. 281 – 284, 332 – 339.

Ludwig F. Weitere Beobachtungen zur Biologie von *Helleborus foetidus* // Bot. Centralbl., Cassel. – 1899. – Bd. 79, 80. – S. 153 – 159, 401 – 413.

Lundstroem A.N. Pflanzenbiologische Studien. 2. Die Anpassungen der Pflanzen an Tiere // Nova Acta Soc. Sc. Ups., Upsala, Ser. 3. – 1887. – Vol. 13, № 2 (cited after Sernander, 1906).

Mabelis A.A. Wood ants wars – the relationship between aggression and predation in the red wood ant (*Formica polyctena* Först.) // Neth. J. Zool. – 1979. – V. 29. – P. 451 – 620.

Malozemova L.A. Interrelation of ants with plants // Ecologia. – 1970. – № 2. – P. 101 – 103 (in Russian).

Malozemova L.A., Koruma N.P. Influence of ants on soil // Ecologia. – 1973. – № 4. – P. 450 – 452 (in Russian).

Mark S., Olesen J.M. Importance of elaiosome size to removal of ant-dispersed seeds // Oecologia. – 1996. – V. 107. – P. 95 – 101.

Marsh A.C. The foraging ecology of two Namib Desert harvester ant species // S. Afr. J. Zool. – 1987. – V. 22. – P. 130 – 136.

Marshall D.L., Beattie A.J., Bollenbacher W.E. Evidence for diglycerides as attractants in an ant-seed interaction // J. Chem. Ecol. – 1979. – V. 5, № 3. – P. 335 – 344.

Martin A.C. The comparative internal morphology of seeds // Amer. Midland Nat. – 1946. – V. 36, № 3. – P. 516 – 660.

Maxwell M. H. Two rapid and simple methods used for the removal of resins from 1.0 µm thick epoxy sections // J. Microsc. – 1978. – V. 112. – P. 253 – 255.

Mesler M.R., Lu K.L. Seed dispersal of *Trillium ovatum* (Liliaceae) in second-growth redwood forests // Amer. J. Bot. – 1983. – V. 70. – P. 1460 – 1467.

Moggridge J.Tr. Harvesting ants and trap-door spiders.– London, 1873 (cited after Sernander, 1906).

Mordak E.V. Wild hyacinthes of the Soviet Union. 1. Morphological and systematical characteristics and their taxonomical significance // Bot. Zh. – 1970. – V. 55, № 9 (cited after Komar, 1978; in Russian).

Mossop M.K. Comparison of seed removal by ants in vegetation on fertile and infertile soils // Austr. J. Ecol. – 1989. – V. 14. – P. 367 – 374.

Muller R.N. The phenology, growth and ecosystem dynamics of *Erythronium americanum* in the northern hardwood forest // Ecol. Monogr. – 1987. – V. 48. – P. 1 – 20.

Nesom G.L. Ant dispersal in *Wedelia hispida* (Heliantheae, Compositae) // Southwest. Natur. – 1981. – V. 26. – P. 5 – 12.

Netolitzky F. Anatomie der Angiospermen-Samen. – Berlin, 1926 (cited after Werker, 1997).

Nicke D.A., Neal T.M. Observations on the foraging behavior of the southern harvester ant, *Pogonomyrmex badius* // Fla. Entomol. – 1972. – V. 55. – P. 65 – 66.

Nordhagen R. Über die Einrollung der Fruchtstiele bei der Gattung *Cyclamen* und ihre biologische Bedeutung // Beih. Bot. Centralbl. – 1932. – Bd. 49. – S. 359 – 395.

Nordhagen R. Remarks on some new or little known myrmecochorous plants from North America and East Asia // Bull. Res. Couns. Isr., Sect. D. Bot. – 1959. – V. 7D. – P. 184 – 201.

Novikov V.P. Diaspores dispersed by ants // Uch. Zap. Len. Gos. Ped. Inst. im. Gertsena. – 1939. – V. 25 (cited after Levina, 1957; in Russian).

O'Dowd D., Gill A.M. Seed dispersal syndromes in Australian *Acacia* // Seed dispersal. – Sydney, Orlando etc.: Academic Press, 1986. – P. 87 – 121.

O'Dowd D.J., Hay M.E. Mutualism between harvester ants and a desert ephemeral: seed escape from rodents // Ecology. – 1980. – V. 61, № 3. – P. 531 – 540.

Oekland F. Kvantitative undersoekelser av de roode skogmaur ernaering // Naturen (Bergen). – 1932. – V. 56. – P. 247 – 252.

Ohara M., Higashi S. Interference by ground beetles with the dispersal by ants of seeds of *Trillium* species (Liliaceae) // J. Ecol. – 1987. – V. 75. – P. 1091 – 1098.

Ohkawara K., Higashi S. Relative importance of ballistic and ant dispersal in two diplochorous *Viola* species (Violaceae) // Oecologia. – 1994. – V. 100. – P. 135 – 140.

Ohkawara K., Higashi S., Ohara M. Effects of ants, ground beetles and the seed-fall patterns on myrmecochory of *Erythronium japonicum* Decne (Liliaceae) // Oecologia. – 1996. – V. 106 – P. 500 – 506.

Ohkawara K., Ohara M., Higashi S. The evolution of ant dispersal in a spring ephemeral *Corydalis ambigua* (Papaveraceae): Timing of seed fall and effects of ants and ground beetles // Ecography. – 1997. – V. 20, № 3. – P. 217 – 223.

Oliveira P.S., Hölldobler B. Orientation and communication in the neotropical ant *Odontomachus bauri* Emery (Hymenoptera, Formicidae, Ponerinae) // Ethology. – 1989. – V. 83. – P. 154 – 166.

Oostermeijer J.G.B. Myrmecochory in *Polygala vulgaris* L., *Luzula campestris* (L.) DC. and *Viola curtisii* Forster in a Dutch dune area // Oecologia. – 1989. – V. 78. – P. 302 – 311.

Pacini E. *Mercurialis annua* L. (Euphorbiaceae) seed interactions with the ant *Messor structor* (Latr.), Hymenoptera: Formicidae // Acta Bot. Neerl. – 1990. – V. 39. – P. 253 – 262.

Passos S., Ferreira S.O. Ant dispersal of *Croton priscus* (Euphorbiaceae) in a tropical semideciduous forest in South Brazil // Biotropica. – 1996. – V. 28, № 4b. – P. 697 – 700.

Petal J. The role of ants in ecosystems // Production ecology of ants and termites / Ed. Brian M.V. – Cambridge: Cambridge Univ. Press, 1978. – P. 293 – 325.

Petit D., Charaf E.O., El Bahja S., Mjid H., Moussadak Z. Les fourmis moissonneues et les composees dans le bassin mediterranen occidental // Bull. Soc. Etude Sci. Natur. Beziers. – 1987 – 1988. – V. 12. – P. 33 – 44.

Pfeiffer A. Die Arillargebielde der Pflanzensamen // Bot. Jahrb. f. Syst. Pflanzengeschichte u. Pflanzengeographie. – 1891. – Bd. 13 (cited after Komar, 1965).

Pijl L.van der. Principles of dispersal in higher plants. – New York, Heidelberg, Berlin: Springer, 1982. – 215 s.

Pudlo R.J., Beattie A.J., Culver D.C. Population consequences of changes in ant-seed mutualism in *Sanguinaria canadensis* // Oecologia. – 1980. – V. 146. – P. 32 – 37.

Radchenko T.A., Malozemova L.A., Filinkova T.N. Interrelations of ants with plants in meadow phytocoenosis // Abstr. Thes. 6th Russ. Myrmecolog. Symposium «Ants and forest defence». – Tartu, 1979. – P. 71 – 75 (in Russian).

Rao A.N., Shamanna S. Rudimentary aril in in *Sanguinaria canadensis* L. // Can. J. Bot. – 1963. – V. 41. – P. 1529 – 1531.

Renner S.S. Seed dispersal // Progr. Bot. – 1987. – P. 413 – 432.

Rice B., Westoby M. Myrmecochory in sclerophyll vegetation of the West Head, New South Wales // Austral. J. Ecol. – 1981. – V. 6. – P. 291 – 298.

Rice B., Westoby M. Evidence against the hypothesis that ant-dispersed seeds reach nutrient-enriched microsites // Ecology. – 1986. – V. 67. – P. 1270 – 1274.

Ridley H.N. Symbiosis of ants and plants // Ann. Bot. – 1910. – V. 24. – P. 457 – 487.

Ridley H. N. The Dispersal of Plants throughout the World.– Kent, 1930.– 744 p.

Roberts J.T., Heithaus E.R. Ants rearrange the vertebrate-generated seed shadow of a neotropical fig tree // Ecology. – 1986. – V. 67, № 4.– P. 1046 – 1051.

Robertson Ch. Seed crests and myrmecophilous dissemination in certain plants // Botan. Gazette, Chicago. – 1897. – V. 23. – P. 288 – 289.

Roth I. Fruits of Angiosperms. – Berlin, Stuttgart: Gebrüder Borntraeger, 1977. – 675 p.

Ruppert K. Dissemination de *Scopolia carinthaica* par les formis // Act. Soc. Bot. Poloniae. – 1893 (cited after Ridley, 1930).

Seidel J. L., Epstein W. W., Davidson D. W. Neotropical ant gardens. 1. Chemical constituents // J. Chem. Ecol. – 1990. – V. 16. – P. 791 – 1816.

Sernander R. Den skandinaviska vegetationens spridningsbiologi. Zur Verbreitungsbiologie der skandinavischen Pflanzenwelt. – 1901 (cited after Sernander, 1906).

Sernander R. Entwurf einer Monographie der europäischen Myrmecochoren // Kungliga Svenska Vetenskapsacademiens Handlingar. – 1906. – Bd. 41, № 7. – S. 1 – 410.

Shea S. R., McCormick J., Portlock C.C. The effect of fires on regeneration of leguminous species in the northern jarrah (*Eucalyptus marginata* Sm.) forest of Western Australia // Austral. J. Ecol. – 1979. – V. 4. – P. 195 – 205.

Sindarovskaya J.R., Gorb E.V., Gorb S.N. Non-myrmecochorous herb *Galium aparine* L. on the mounds of *Formica polyctena* Fabr.: probable benefits of the occasional relation // Zh. Obshch. Biol. – 1997. – V. 58, № 6. – P. 76 – 80 (in Russian).

Singh D. Structure and development of ovule and seed of *Viola tricolor* and *Ionidium suffruticosum* Ging. // J. Ind. Bot. Soc. – 1963. – V. 42, № 3. (cited after Tahtadzhjan, 1992).

Singh B. Development and structure of angiosperm seed. I. Review of Indian work // Bull. Nat. Bot. Gard. (Lucknow). – 1964. – V. 89. – P. 1 – 115.

Singh D. Passifloraceae, Violaceae, Caricaceae // Bull. Ind. Nat. Sci. Acad. – 1970. – V. 41 (cited after Tahtadzhjan, 1992).

Skidmore B.A., Heithaus E.R. Lipid cues for seed-carrying by ants in *Hepatica americana* // J. Chem. Ecol. – 1988. – V. 14. – P. 2185 – 2196.

Smith J.M.B. An example of ant-assisted plant invasion // Austral. J. Ecol. – 1989. – V. 14, № 2. – P. 247 – 250.

Spurr A.R. A low-viscosity epoxy resin embedding medium for electron microscopy // J. Ultrastruct. Res. – 1969. – V. 26. – P. 31 – 43.

Sykes W.H. Descriptions of new species of indian ants // London, Transact. Entomol. Soc., London. – 1834 (cited after Sernander, 1906).

Szemes G. Zur Entwicklung des Elaiosoms von *Chelidonium majus* // Wien. Bot. Z. – 1943. – Bd. 92. – S. 215 – 219.

Targonsky P.N. Myrmecochorous plants of Zhytomyr District // Ukr. Bot. Zh. – 1995. – V. 52, № 1 – P. 68 – 71 (in Ukrainian).

Tahtadzhjan A.L. Principles of evolutionary morphology of angiosperms. – Leningrad: Nauka, 1964. – 236 p. (in Russian).

Tahtadzhjan A.L. (ed.) Plants' life. – Moskwa: Prosveshchenije, 1980. – V. 5 (1). – 430 p. (in Russian).

Tahtadzhjan A.L. (ed.) Plants' life. – Moskwa: Prosveshchenije, 1981. – V. 5 (2). – 511 p. (in Russian).

Tahtadzhjan A.L. (ed.) Plants' life. – Moskwa: Prosveshchenije, 1982. – V. 6. – 542 p. (in Russian).

Tahtadzhjan A.L. (ed.) Comparative anatomy of seeds. – Leningrad: Nauka, 1985. – V. 1: Monocotyledones. – 317 p. (in Russian).

Tahtadzhjan A.L. (ed.) Comparative anatomy of seeds. – Leningrad: Nauka, 1988. – V. 2: Dicotyledons. Magnoliidae, Ranunculidae. – 257 p. (in Russian).

Tahtadzhjan A.L. (ed.) Comparative anatomy of seeds. – Leningrad: Nauka, 1991. – V. 3: Dicotyledons. Caryophyllidae – Dilleniidae. – 240 p. (in Russian).

Tahtadzhjan A.L. (ed.) Comparative anatomy of seeds. – Leningrad: Nauka, 1992. – V. 4: Dicotyledons. Dilleniidae. – 432 p. (in Russian).

Thompson S.N. A review and comparative characterization of the fatty acid compositions of seven insect orders // Compar. Biochem. and Physiol. – 1973. – V. 45. – P. 467 – 482.

Thompson N. Elaiosomes and fleshy fruits: phenology and selection pressures for ant-dispersed seeds // Amer. Nat. – 1981. – V. 117. – P. 104 – 108.

Traniello J.F.A. Foraging strategies of ants // Annu. Rev. Entomol. – 1989. – V. 34. – P. 191 – 210.

Turnbull C.L., Culver D.C. The timing of seed dispersal in *Viola nuttallii*: attraction of dispersers and avoidance of predators // Oecologia. – 1983. – V. 59. – P. 360 – 365.

Ulbrich E. Deutsche Myrmekochoren, Beobachtungen über die Verbreitung heimische Pflansen durch Ameisen. // Leipzig, Berlin: Verlag von Theodor Fisher, 1919.– 56 S.

Ulbrich E. Biologie der Früchte und Samen (Karpobiologie). Berlin: Springer-Verlag, 1928. – 230 S.

Uphof J.C. T. Ecological relations of plants and termites // Bot. Rev. – 1942.– V. 8.– P. 563 – 598.

Vinson S.B., Thompson J.L., Green N.B. Phagostimulants for the imported fire ant *Solenopsis saevissima richerti* // J. Insect Physiol. – 1967. – V. 13. – P. 1729 – 1736.

Wagner D. The influence of ant nests on *Acacia* seed production, herbivory and soil nutrients // J. Ecol. 1997. – V. 85. – P. 83 – 93.

Wein G.R., Pickett S.T.A. Dispersal, establishment and survivorship of a cohort of *Erythronium americanum* // Bull. Torrey Bot. Club. – 1989. – V. 116. – P. 240 – 246.

Wellenstein G. Zur Ernährungsbiologie der Roten Waldameise (*Formica rufa* L.) // Z. Pflanzenkr. – 1952. – V. 59. – P. 430 – 451.

Werker E. Seed anatomy. – Berlin, Stuttgart: Gebrüder Borntraeger, 1997. – 424 p.

Westoby M., French K., Hughes L., Rice B., Rodgerson L. Why do more plant species use ants for dispersal on infertile compared with fertile soils? // Austral. J. Ecol. – 1991a. – V. 16, № 4. – P. 445 – 455.

Westoby M., Hughes L., Rice B.L. Seed dispersal by ants; comparing infertile with fertile soils // Ant-plant interactions / Eds. Huxley C.R., Cutler D.F. – Oxford, New York., Tokyo: Oxford Univ. Press, 1991b. – P. 434 – 447.

Westoby M., Rice B. A note on combining two methods of dispersal-for-distance // Austral. J. Ecol. – 1981. – V. 6. – P. 1 – 4.

Westoby M., Rice B., Howell J. Seed size and plant growth form as factors in dispersal spectra // Ecology. – 1990. – V. 71, № 4. – P. 1307 – 1315.

Westoby M., Rice B., Shelley J.M., Haig D., Kohen J.L. Plants' use of ants for dispersal at West Head, New South Wales // Ant-plant interactions in Australia / Ed. Buckley R.C. – The Hague: Dr. W. Junk Publishers, 1982. – P. 75 – 87.

Wheeler Wm.M., Bequart J., Bailey J.W. Ants of the American Museum Congo Expedition. A contribution to the myrmecology of Africa. – 1921 – 1922 (cited after Ulbrich, 1928).

Wilson E.O. The Insect Societies. – Cambridge, Massachusetts: The Belknap Press of Harvard University Press, 1971. – 548 p.

Wilson E.O., Durlach N.I., Roth L.M. Chemical releasers of necrophoric behavior in ants // Psyche. – 1958. – V. 65. – P. 108 – 114.

Windsor D.M., Jolivet P. Aspects of the morphology and ecology of two Panamanian ant-plants, *Hoffmannia vesciculifera* (Rubiaceae) and *Besleria formicaria* (Gesneriaceae) // J. Tropical Ecology. – 1996. – V. 12, № 6. – P. 835 – 842.

Woodell R.J, King T.J. The influence of mound-building ants on British lowland vegetation // Ant-plant interactions / Eds. Huxley C.R., Cutler D.F. – Oxford, New York, Tokyo: Oxford Univ. Press, 1991. – P. 521 – 536.c

Yu D.W. The structural role of epiphytes in ant gardens // Biotropica. – 1994. – V. 26. – P. 222 – 226.

Appendix

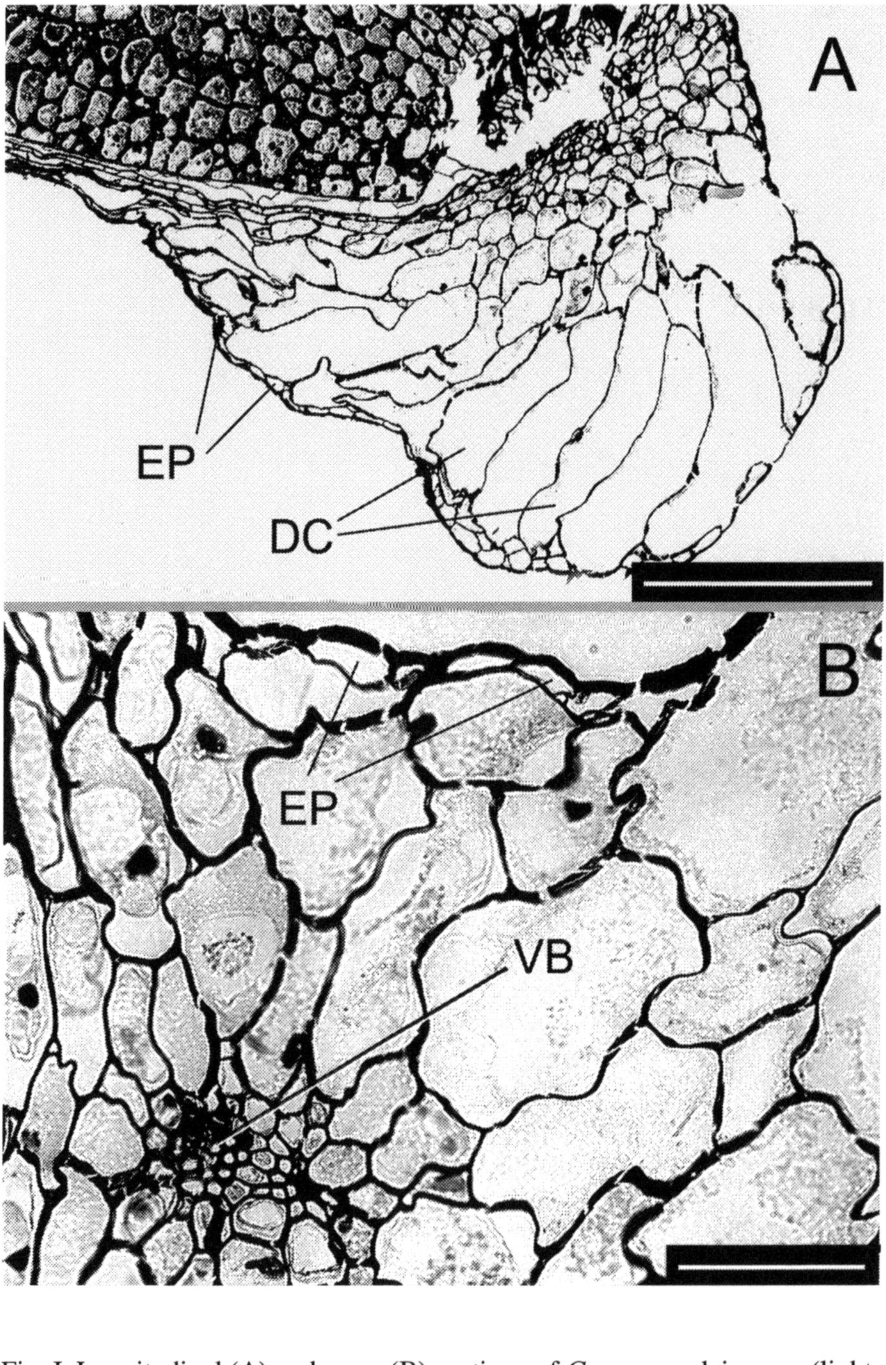

Fig. I. Longitudinal (A) and cross (B) sections of *Gagea* sp. elaiosome (light microscope, toluidine blue stain). DC, distinct cortex of elaiosome; EP, epidermis; VB, vascular bundle. Scale bar: A – 500 μm, B – 100 μm.

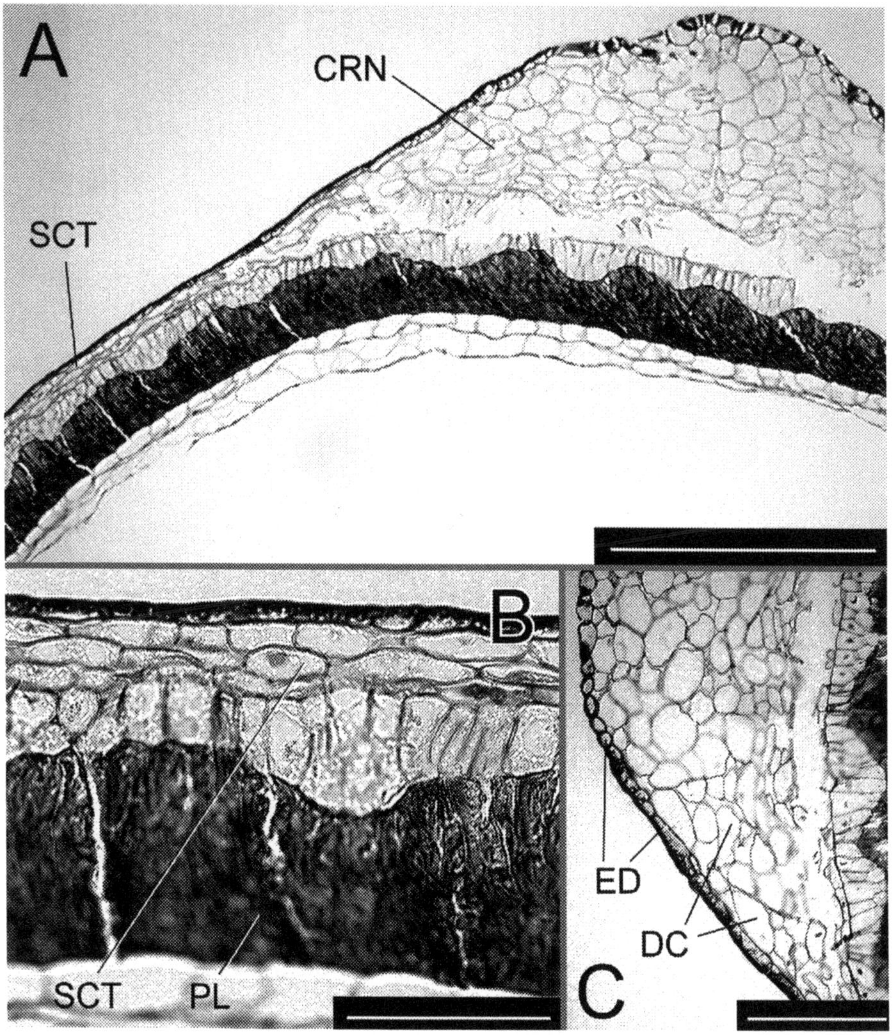

Fig. II. Longitudinal sections of *Mercurialis perennis* sarcotesta and caruncle (light microscope, toluidine blue stain). A. Combination of sarcotesta and caruncle. B. Sarcotesta. C. Caruncle. CRN, caruncle; DC, distinct cortex of elaiosome; ED, epidermis; PL, protective layer; SCT, sarcotesta. Scale bar: A – 500 μm, B – 100 μm, C – 200 μm.

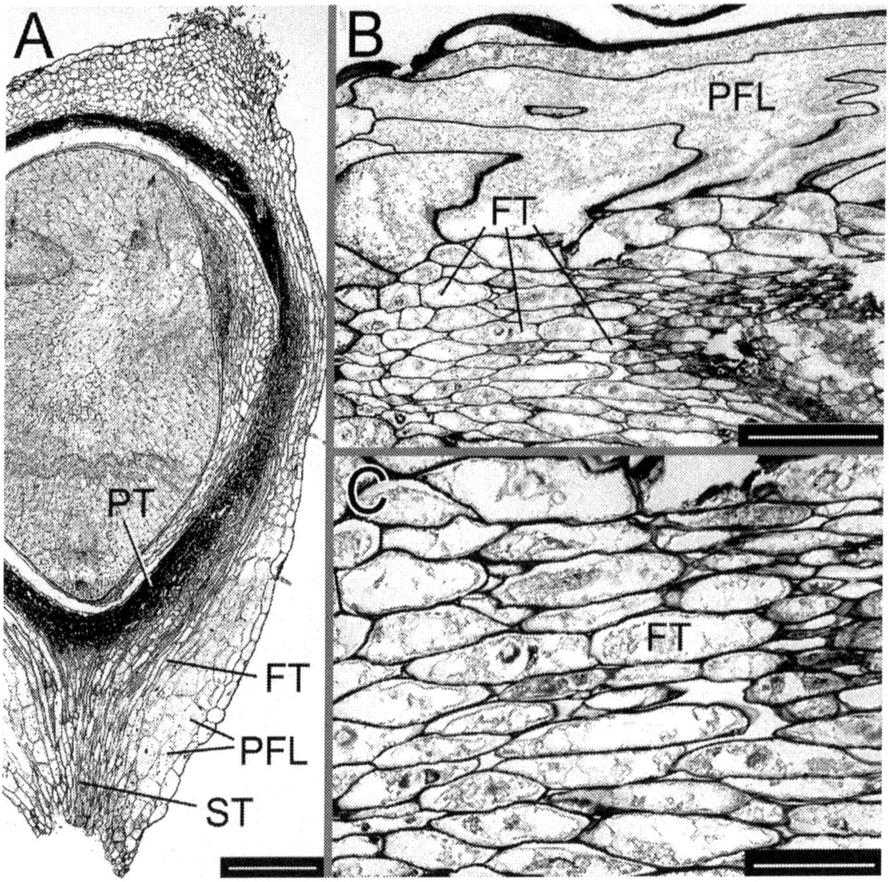

Fig. III. Longitudinal sections of *Ficaria verna* fruit (A) and *Lamium maculatum* elaiosome (A, B) (light microscope, toluidine blue stain). A. General view. B. Peripheral part. C. Central part. FT, fundamental tissue; PFL, peripheral layer; PT, protective tissue; ST, supporting tissue (vascular bundle). Scale bar: A – 500 μm, B – 100 μm, C – 50 μm.

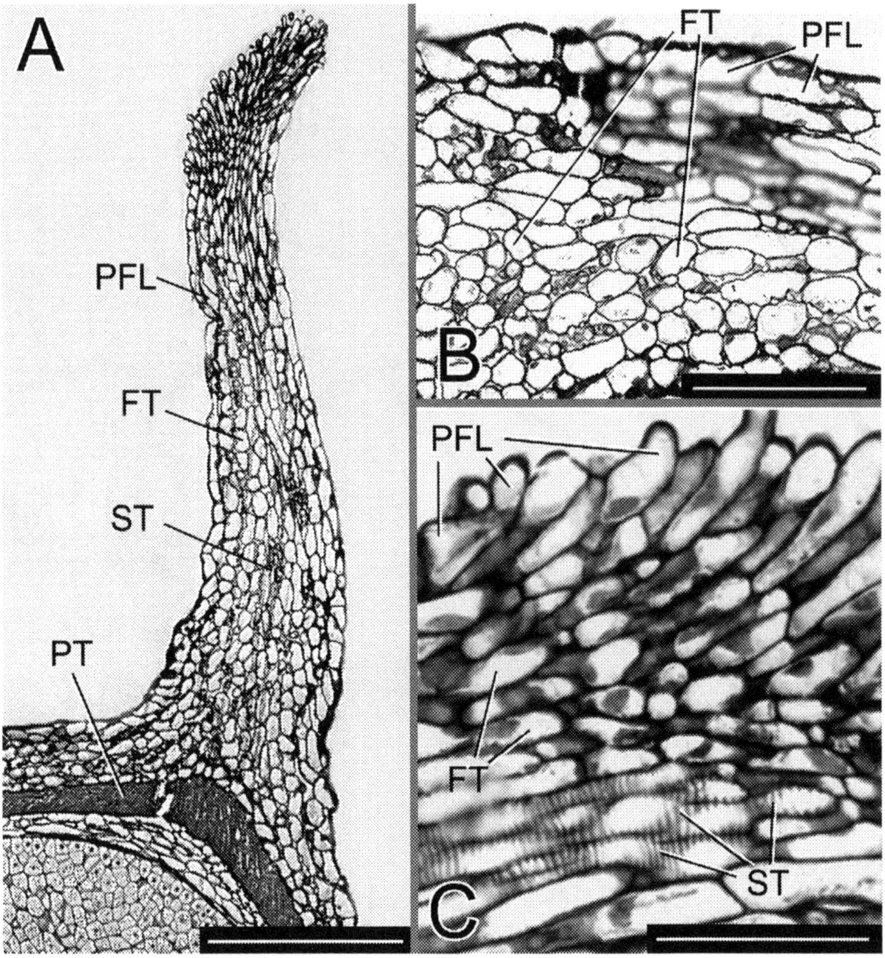

Fig. IV. Longitudinal sections of *Anemone ranunculoides* elaiosome (light microscope, toluidine blue stain). A. General view. B. Central part. C. Apical part. FT, fundamental tissue; PFL, peripheral layer; PT, protective tissue; ST, supporting tissue (vascular bundle). Scale bar: A – 500 μm, B – 200 μm, C – 100 μm.

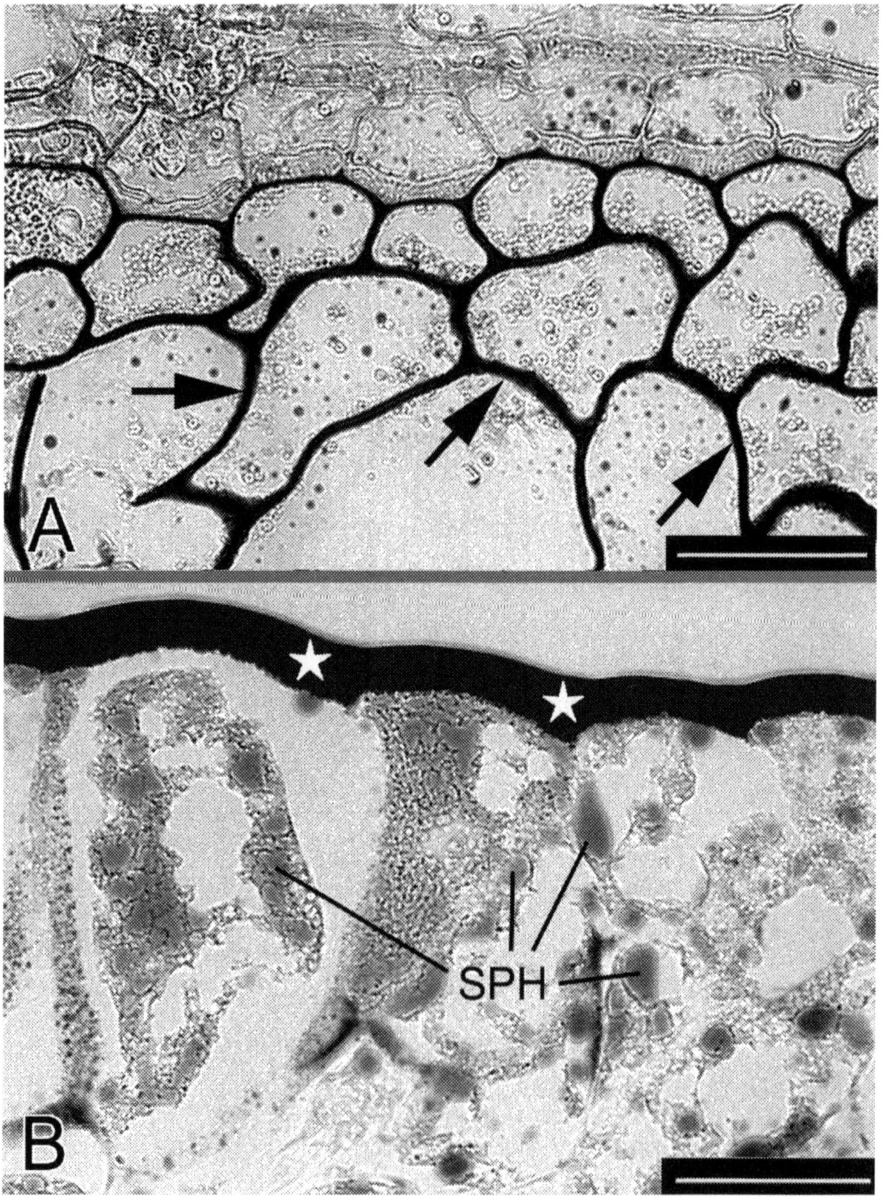

Fig. V. Elaiosome cells of *Chelidonium majus* (A) and *Viola reichenbachiana* (B) (light microscope, Sudan Black stain). Lipids are connected with cell walls (arrows), appear in spherosomes (SPH) or in an external layer of the elaiosome (stars). Scale bar: 50 μm.

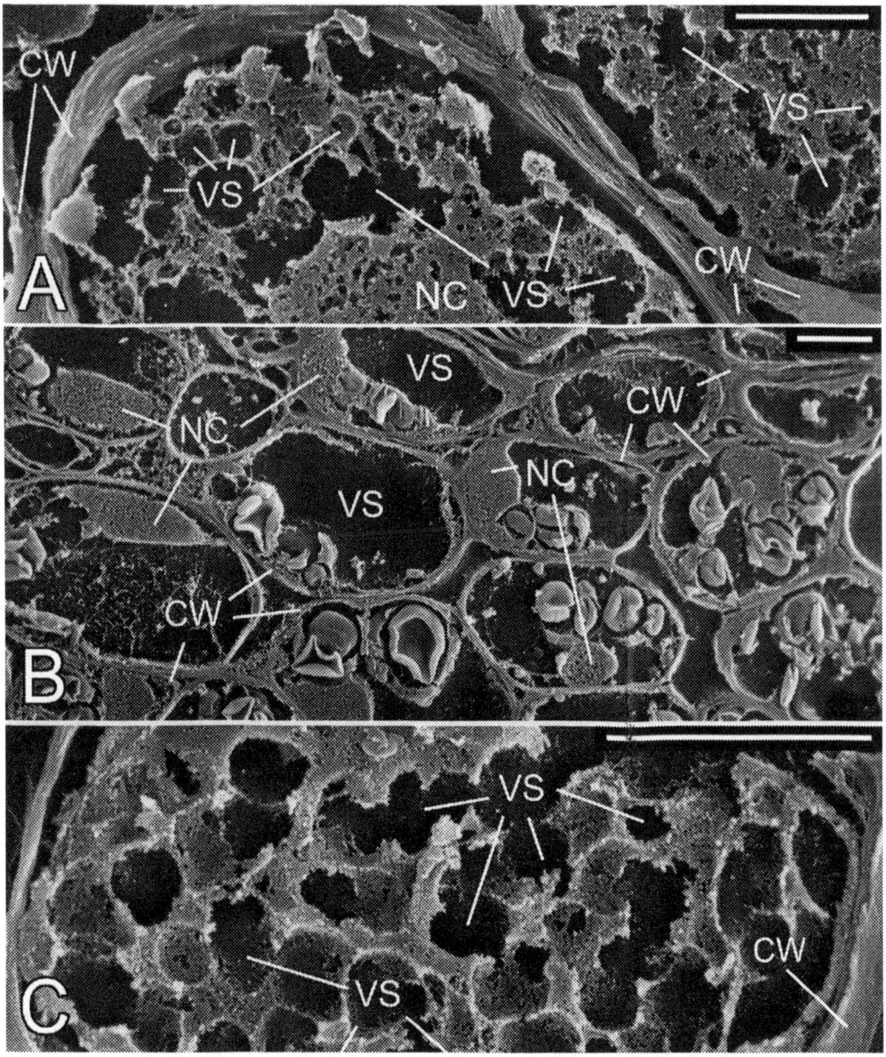

Fig. VI. Semi-thin sections of elaiosome cells of *Pulmonaria obscura* (A), *Lamium maculatum* (B) and *Mercurialis perennis* (C) in scanning electron microscope (after Maxwell method (Maxwell, 1978)). CW, cell walls; NC, nuclei; VS, places of lipid localisation. Scale bar: 5 μm.

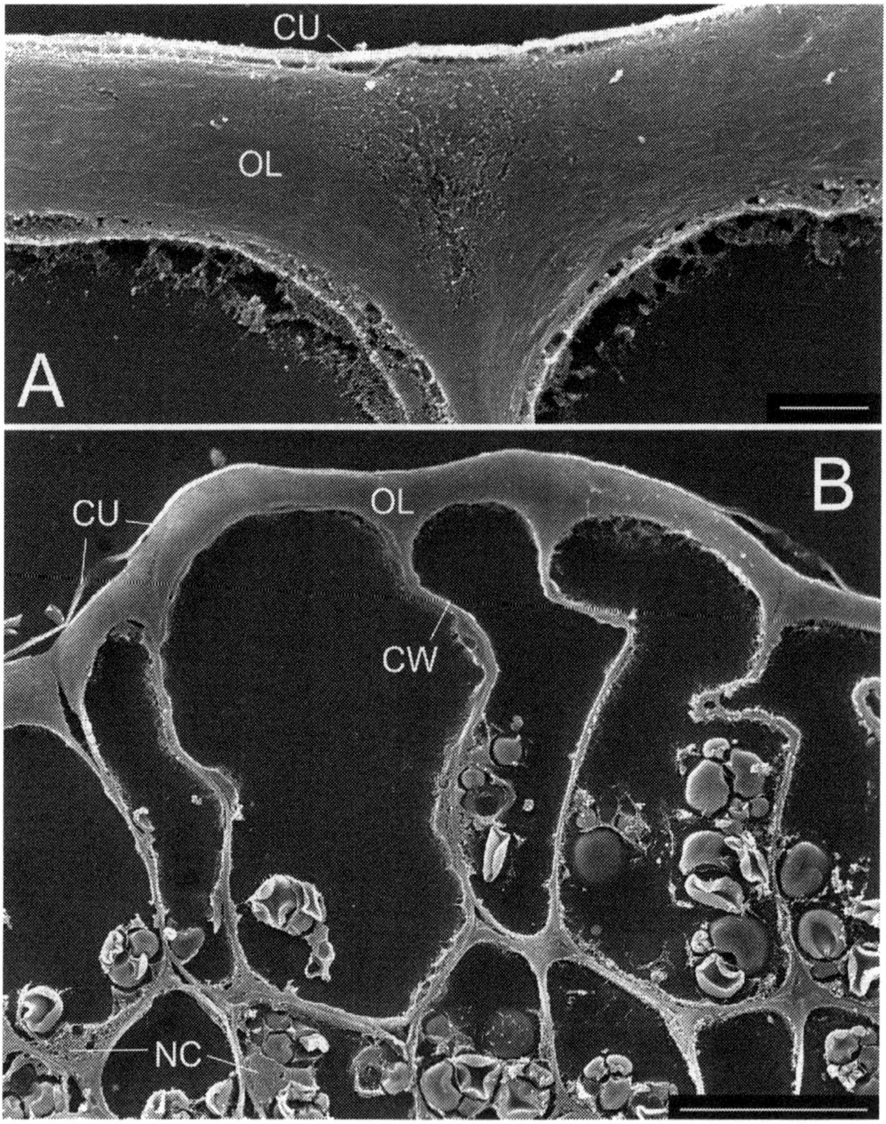

Fig. VII. Semi-thin sections of elaiosome cells of *Lamium maculatum* in scanning electron microscope (after Maxwell method (Maxwell, 1978)). CU, cuticle; CW, cell walls; NC, nuclei; OL, external lipid layer. Scale bar: A – 2 μm, B – 20 μm.

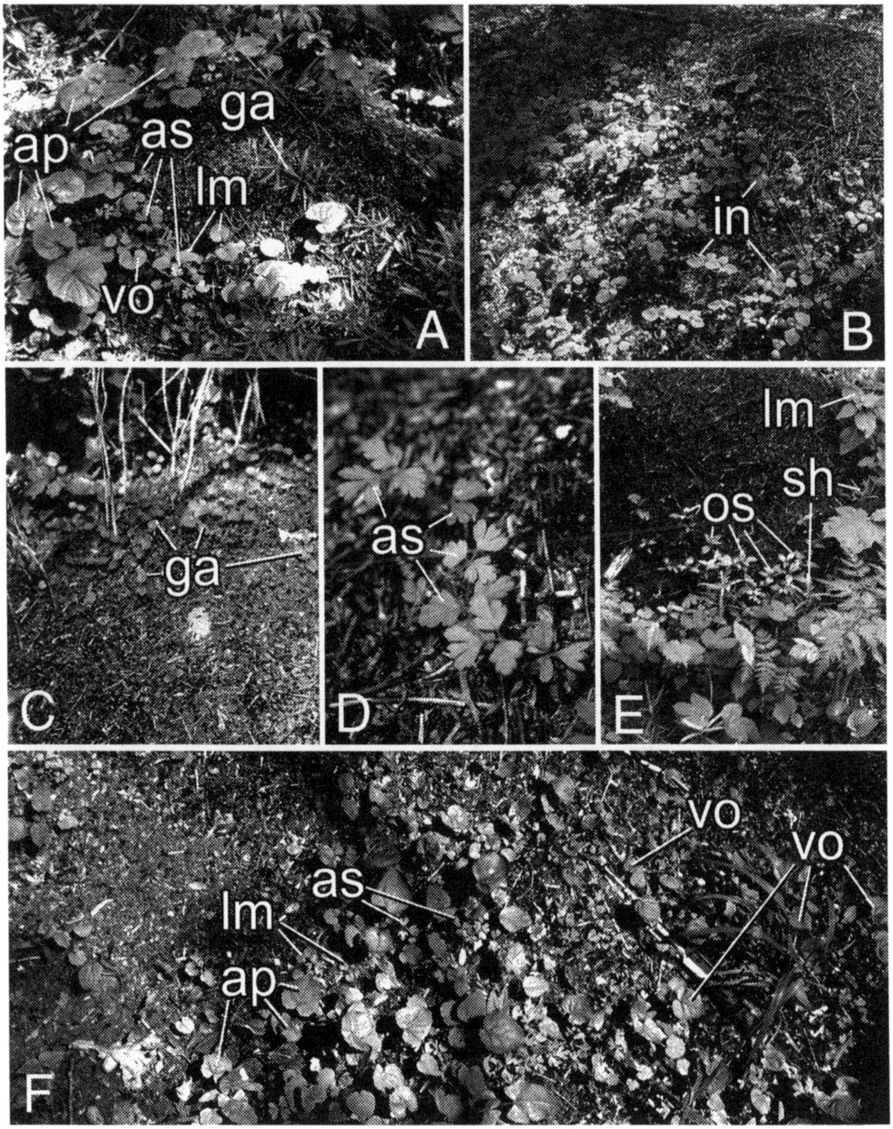

Fig. VIII. Vegetation of *Formica polyctena* nest mounds. ap, *Alliaria petiolata*; as, *Anthriscus sylvestris*; ga, *Galium aparine*; in, *Impatiens noli-tangere*; lm, *Lamium maculatum*; os, *Omphalodes scorpioides*; sh, *Stellaria holostea*; vo, *Viola odorata*.

Fig. IX. A. Study site. B. Guilds of myrmecochores in the deciduous forest. C. Germination success of seeds of *Viola odorata* collected from garbage dump on the territory borders of *Formica polyctena* ants (n=238) and seeds taken directly from the fruit (n=238). D. Diaspores collected from soil samples from various forest microsites. E. Experimental set-up in study on seedling survival of *Coridalis bulbosa*. F. Typical set-up for «cafeteria» experiments. ae, *Asarum europaeum*; lm, *Lamium maculatum*; mp, mature plants; po, *Pulmonaria obscura*; sd, seedlings; sds, seeds; tb, wooden table; vh, *V. hirta*; vo, *V. odorata*; wk, ant workers.

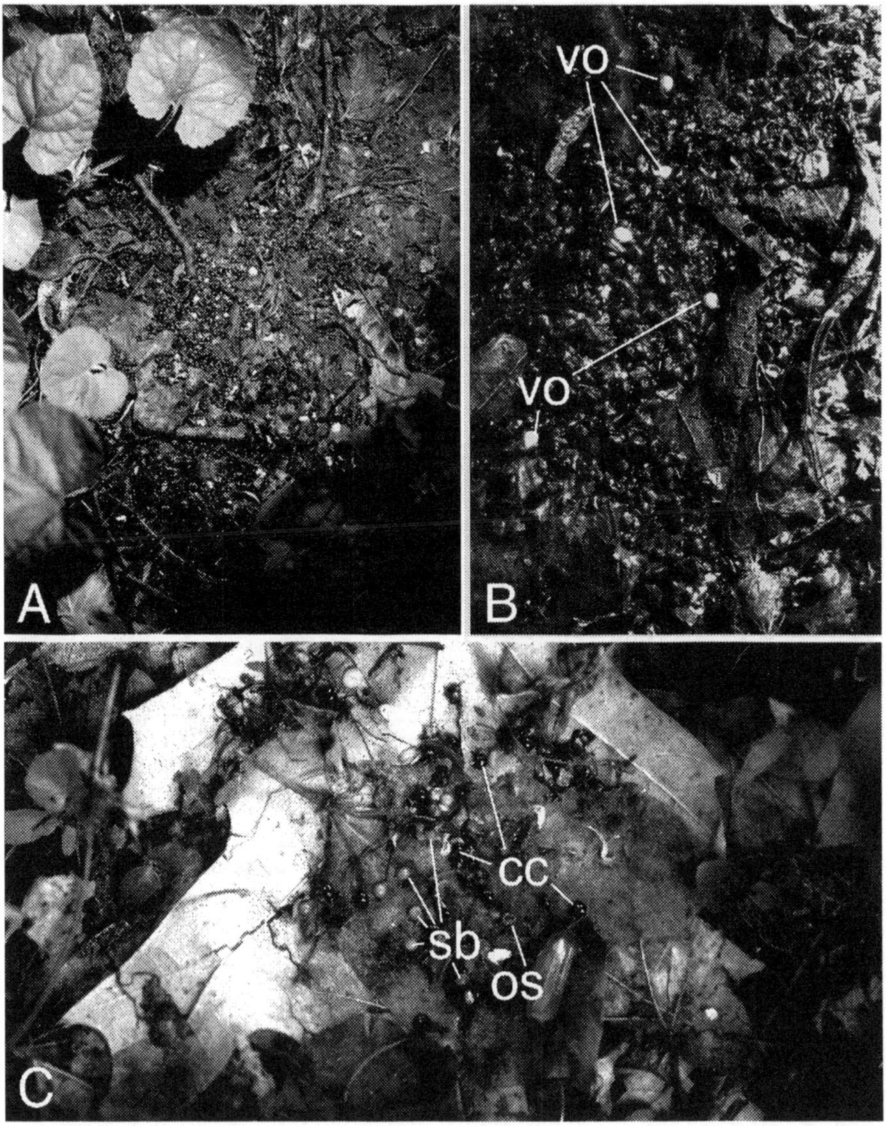

Fig. X. Garbage dump of *Formica polyctena* ants on the new border between territories of *Lasius fuliginosus* and *Formica polyctena* in late Summer (A, B) and late Spring (C). Note presence of diaspores among insect remains. cc, *Corydalis cava*; os, *Omphalodes scorpioides*; sb, *Scilla bifolia*; vo, *Viola odorata*.

Fig. XI. Association of seedlings and juvenile plants on the old border between territories of *Lasius fuliginosus* and *Formica polyctena*. ae, *Asarum europaeum*; cb, *Corydalis bulbosa*; lm, *Lamium maculatum*; os, *Omphalodes scorpioides*; sh, *Stellaria holostea*; vo, *Viola odorata*.

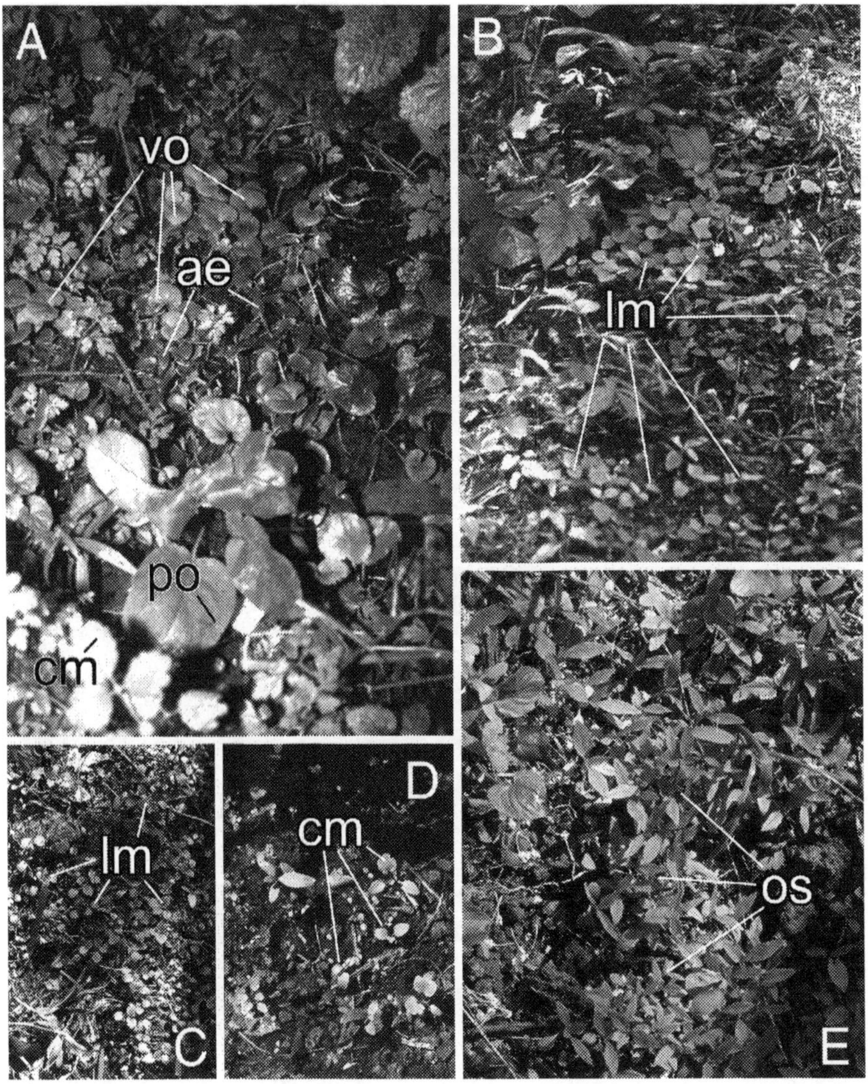

Fig. XII. Ant-related associations of plants in the deciduous forest. A. Border between *Formica polyctena* and *Lasius fuliginosus*. B, C. A high concentration of *Lamium maculatum* in surroundings of *L. fuliginosus* nest (B) and on the border between *F. polyctena* and *L. fuliginosus* territories (C). D. A high concentration of *Chelidonium majus* on the *L. niger* nest. E. A high concentration of *Omphalodes scorpioides* on the border between *F. polyctena* and *L. fuliginosus* territories. ae, *Asarum europaeum*; cm, *C. majus*; lm, *L. maculatum*; os, *O. scorpioides*; po, *Pulmonaria obscura*; vo, *Viola odorata.*

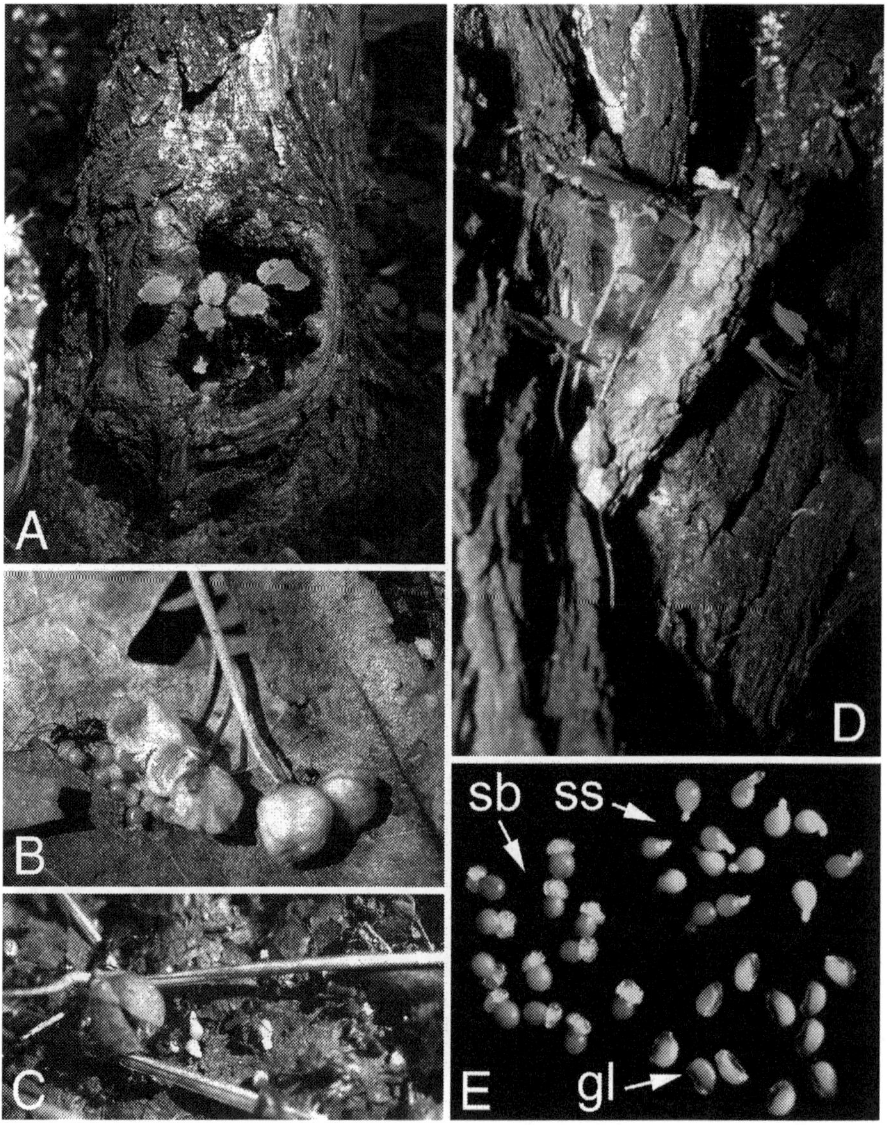

Fig. XIII. Fate of seeds of myrmecochorous plants. A. Juvenile plants of *Lamium maculatum* growing in a *Myrmica rubra* nest located in the hollow of a tree. B. *Formica polyctena* worker removing seeds from the opened fruit of *Scilla bifolia*. C. *F. polyctena* workers removing seeds from the opened fruit of *V. odorata*. D. Juvenile plant of *Chelidonium majus* growing in a *Lasius niger* nest located on the tree bark. E. Elaiosome-bearing seeds of early-spring myrmecochores: *Gagea lutea* (gl), *S. bifolia* (sb), *S. sibirica* (ss).

Fig. XIV. Fruits and seeds of myrmecochorous plants from the deciduous forest. A. *Viola mirabilis*. B. *Chelidonium majus*. C, E. *Scilla sibirica*. D. *V. odorata*. F. *Pulmonaria obscura*.

Fig. XV. Fruits and seeds of myrmecochorous plants from the deciduous forest. A. *Anemone ranunculoides*. B. *Mercurialis perennis*. C. *Corydalis cava*. D. *Omphalodes scorpioides*. E. *Asarum europaeum*. F. *Ficaria verna*.

Index

B

C

D

E

F

G

H

I

K

L

M

N

O

P

Q

R

S

T

U

V

Z